NATURAL MAN

Robert Henry Welker

NATURAL MAN

The Life of William Beebe

INDIANA UNIVERSITY PRESS
Bloomington & London

Published in Canada by Fitzhenry & Whiteside Limited,
Don Mills, Ontario

Manufactured in the United States of America

Library of Congress Cataloging in Publication Data

Welker, Robert Henry.
Natural man : the life of William Beebe.

Bibliography
Includes index.
1. Beebe, Charles William, 1877–1962. I. Title.
QL31.B37W44 1975 591'.092'4 [B] 74-22834
ISBN 0-253-33975-8 75 76 77 78 79 1 2 3 4 5

To Cynthia

CONTENTS

ILLUSTRATIONS

PREFACE

Famous though he was in his own time, William Beebe chose never to write an autobiography. His rivals for acclaim among the naturalists, Frank M. Chapman, T. Gilbert Pearson, and Ernest Thompson Seton, all had done so; but Beebe held aloof. Furthermore, as reported by a friend of half a century, "Dr. Beebe in his will left his papers . . . with the understanding that they would not be made available to anyone. He also frequently said that he did not want anybody to write his biography."

Hence a natural question: why such reticence? That some things are better left undisclosed is true of any life, and certainly of Beebe's, as the present study will suggest. But the scandalmonger plays at best an ignoble part, and the gossip a paltry one. One doubts that a man of Beebe's abundant self-esteem felt any major threat from such as these. So perhaps there were more important things which troubled him, things intellectual and professional rather than personal. In his fifties William Beebe knew that he had become America's best-known naturalist; but he also knew that natural science has a way of reckoning the accounts of all but the greatest contributors, and pressing on. Any biography is a reckoning, and something to be deferred if not escaped. Beebe well understood both the tests of fame and the more rigorous criteria of achievement. Perhaps he did not wish a judgment rendered, whether by himself or by another.

But he lived a long life, and toward the end there were already younger naturalists who, in George Sarton's metaphor, were standing on Beebe's shoulder, the farther to see. There were others who judged his work and found it wanting; and finally there were those who passed him by with scarcely a glance. In short, the reckoning had begun. But still no autobiography was forthcoming, and at last there came the effort from beyond the grave to discourage any life others might undertake to write.

Whether his name as a scientist would endure or fade, Beebe surely perceived, was beyond effective manipulation or useful argument. He had offered in cold print his data and the formulations and speculations based upon them. Science declares by its very name that it knows what it is about, moving forward with cool assurance, accepting or rejecting, but in any event superseding the work of yesterday, that which has served its purpose by engendering the work of today. Although its claim of objectivity and progress may be false, it is no less inescapable for that; and Beebe as he grew older could scarcely hope by personal apologia to appeal decisions made or in the making. Nor, as a proud man, was he likely to enter the kind of special plea many an autobiography proves itself to be.

Yet it must also have occurred to Beebe that after a fashion he had written book after book in the autobiographical mode—indeed, by the end of his life he had written twenty of them. No American naturalist of recent times has given us so copious and detailed a record of his work, his travels, the challenges he faced and his responses to them. These books were not merely of scientific reportage, but of scientific adventure; more important, they were creative works, informed by an intelligence of extraordinary range and a style and consciousness highly literary. In the sense that all good creative writing is revealing of the writer, these books of William Beebe give us the man. But surely not all at once, nor always fully, nor with complete candor; there is much to seek beneath the surface of his art, and much that has been debarred and must be discovered elsewhere. Hence I have conceived this book not as a conventional biography but as a critical inquiry and analysis. The man who emerges from such a study is still the public Beebe, the joyful partisan of wild nature, the intrepid explorer, the devotee and celebrant of life. But he is more: a man of greater complexity, and even of mystery, than he chose in his own time to reveal.

ACKNOWLEDGMENTS

Although one's intellectual and creative creditors are of the kindest sort, never calling in their loans, the debts I owe them cannot pass unacknowledged. Dr. Dean Amadon, the distinguished ornithologist and author at the American Museum of Natural History, had known me only through correspondence, yet on the occasion of our first meeting offered a free and unlimited access to the museum's private library, and thus at a crucial time afforded me research materials of a depth and range available nowhere else. Here in the quiet stacks were gathered half a century and more of primary sources and contemporary records, everything from Beebe's rare *Monograph of the Pheasants* to his annual reports and fugitive pieces, and many essential notes and comments by his colleagues. To Dr. Amadon's unsolicited kindness and the rich scholarly resources that proceeded from it, the word indispensable exactly applies.

Clearly the same is true of the response I received from A. E. Hill, the Canadian physician who attended William Beebe in his last years. In the face of contradictory reports of Beebe's decline and death, only Dr. Hill could supply the needed professional data. Reached in Vancouver through a mutual Trinidadian friend, Mrs. Harry Ross, he did just that, fully, generously, and with persuasive forthrightness. The final datum in this melancholy search, the official death certificate, was obtained after considerable bureaucratic contention by another good Trinidadian, Mrs. Frank Nothnagel.

Among the friendly and informative people of Trinidad and Tobago, I must also include Ian Lambie, secretary of the Trinidad Field Naturalists' Club; O. Marcus Buchanan, curator of the William Beebe Tropical Research Station at Simla; Captain and Mrs. Milton Miles; Captain John Grell; Dr. and Mrs. Hilton Clarke, Kurt and Wanda Nothnagel, Jean Morshead, Errol Newbold, and the late Cecile Brinkley.

From time to time, friends and colleagues in the life sciences have offered useful comments, whether instructive, corrective, or even admonitory: Rosi Kuerti, Marcus Singer, Mickey Sloss, Ted Voneida, Eugene Perrin, and Ralph D. Morris. Similarly I have received aid in physical calculations from Philip Taylor, and essential research data from Sarah Taylor, kindly gathered on a visit to England. An old acquaintance at the American Museum of Natural History, senior entomologist John C. Pallister, offered a genial welcome and some interesting observations on Beebe as well.

Two good attorneys have helped me in diverse ways, Samuel Handelman with the book's title, and Christian Waag with Beebe's divorce records. Elmer Newman, something of a naturalist as well as an outstanding reference librarian, has frequently suggested sources for my work which I might otherwise have slighted. And now that the book is done, I am gratefully aware how many improvements have been wrought by editors as supportive and yet as rigorous as Adam Horvath and Amy Perry.

Having read many an affectionate acknowledgment offered by male writers to their constantly helpful and valiantly patient wives, I supposed from the beginning that I would follow suit with my wife Cynthia—but I had no reason to suspect in what special terms. Although I had studied the lives of many naturalists in prior years, I had written only synoptic essays for biographical reference works. The intense personal engagement that can develop between a biographer and his subject was nearly unknown to me. Robert E. Schofield, himself acclaimed for his work on the life of Joseph Priestley, had offered the cautionary words of Macauley about avoiding "the *Lues Boswelliana,* or disease of admiration." By this friendly warning I thought myself forearmed, even with a history of enthusiasm for William Beebe dating from boyhood. But I was in no way prepared to confront a far more crippling disability, that of alienation. It came nevertheless, in the midst of a period of concentrated writing lasting many months. The symptoms of the affliction intensified from irritation with Beebe's small fakeries to exasperation with his cavalier manipulations of events to anger and at length despair over his radical distortion of the later years with Mary, his first wife. At last I told Cynthia that I felt I could not go on.

Of course she knew that I would not turn to mere debunking, nor assume the stance of cool academic distaste found in too many biographies we both had read. If a basic moral empathy could not be restored, the enterprise was ended. This in particular she understood. She knew as well that I had come to Beebe too simply, seeing him only in the light. But it was a man's history and not a hero's I had set myself to write, and no man of worth has only light to show. When at Cynthia's gentle behest I took up the track again, it was to find the hero diminished nearly to common estate, but the man somberly enhanced by the very shadows about him.

NATURAL MAN

Among the world's great naturalists, few have had a life as adventure-packed as William Beebe. —ROGER TORY PETERSON

:1 Father of the Man

> One of my earliest and most cruel disillusionments came during a season of grammar school wrestling with geography, when I left my pink natal state of New York and, in the course of a short trip discovered to my disgust that New Jersey was not blue, nor Pennsylvania even scarlet.
>
> *Pheasant Jungles*

Thus whimsically did William Beebe describe the experience of leaving Brooklyn, where he was born on July 29, 1877, and moving to East Orange, New Jersey, in his early grade school days.[1] However playful the words recounting his childish disappointment, Beebe, looking back, no doubt knew that the move was a decisive one. Any change from one kind of environment to another is important in the life of a child, and for a boy destined to become a famous naturalist this was especially so. East Orange in those days was a town of perhaps ten thousand people in what was still a rural setting; Brooklyn had grown into a city of more than half a million.[2] Though in colonial times "Brookland Parish" still had a great deal of countryside along the "Road to Flatbush," nearly all of it had vanished, leaving Prospect Park and a cemetery or two as green islands in the midst of a burgeoning metropolis. In such surroundings young Beebe might still have discovered and pursued his naturalist's calling, but surely to move from a great urban center to a modest New Jersey town was a favorable augury.

In William's infancy the family had lived in substantial row houses of stone on St. Marks Avenue and Carlton Avenue, only a block or two from Brooklyn's main thoroughfare, Flatbush Avenue. In these dwellings also lived the boy's grandfather and his uncle

Clarence, but the household broke up when the grandfather died in 1880 and Clarence Beebe moved away. For another few years the family lived elsewhere in Brooklyn, and then occupied at least three houses in succession in East Orange during the latter 1880s before settling permanently at 73 Ashland Avenue.

This place, fondly recalled in years to come, was a proper late Victorian frame house, high and narrow, with a full third story lighted by five dormers and two windows set in the hooded front gable, fringed by decorative wooden scrollwork. Slate of variegated pattern covered the steep hipped roof, and, as at numbers 71 and 69, a one-story roofed porch extended across the front, shadowing the bay window of the parlor and the tall, deeply paneled, double front door. The rear portion of the long but narrow lot was taken up by a back yard and small garden and the carriage house. Maples and horse chestnuts had been planted in the tree lawn facing the street. New or nearly so when the Beebes moved in, this home was an entirely respectable dwelling for Charles Beebe, dealer in paper, with offices at 132 Nassau Street in lower Manhattan.

And one may surmise that it was also a house that was good to grow up in for an active boy like Charles William, the family's only child. (A second son, John Younglove, had been born in Brooklyn in July, 1881, but died fifteen months later.) Here is one of Beebe's affectionate recollections: "For a year and a half when I was a boy, I used to climb out of a dormer window every late afternoon . . . and . . . write a description of the sunset."[3] And another: "I can well remember the time when, on cold wintry mornings, an alarm clock and my mother's impatient voice were but silence compared with the caw of a passing crow."[4] Devoted interest, persistence, aspiration, even a certain boyish eccentricity are suggested here—with a margin to life wide enough to allow them. Such earnest endeavor to no immediate profit was scarcely permitted the usual American boy brought up on the sturdy material homilies of the McGuffey Readers and Horatio Alger, Jr. Obviously this was no usual boy, nor was his upbringing cast in the common end-of-the-century mold.

Yet Beebe's mention of his mother's impatient voice hints at the maternal taskmaster. More explicitly, one writer has called Henrietta Beebe "a woman of intense drive [who was] determined to see that her son should have the benefit of all available aid in the furtherance of his choice of professions and made sure that he met the leaders in the field."[5] That she was ambitious for her son is plain, and East Orange was within easy distance by rail of New

York and its scientific institutions, in particular the American Musem of Natural History. Once it was clear that William (he dropped the "Charles" as early as high school) was serious about his career in nature study, his mother was a force to be reckoned with, as more than one naturalist of the day discovered. Before she died in 1925 she knew the satisfaction of having her son recognized for his work as a scientist and writer by no less a personage than Theodore Roosevelt.

Her husband Charles comes through less sharply. His marriage to Henrietta Marie Younglove of Glens Falls, New York took place in 1875. At the time of William's birth Charles was employed as a clerk in his father's paper business, and he continued to work in this and allied fields through his seventy-ninth year, dying in 1931. It appears his principal civic interest was the New York National Guard; he served for many years as a lieutenant in the Gatling Battery in Brooklyn. He and his twin brother Clarence were born in 1852 in New York City, sons of the merchant Roderick Beebe (1816–1880), the third in the Beebe line to bear that given name. Roderick his father had died in 1831 "of overexertion at a ploughing bee, weighed over 200 pounds, of Chatham, N.Y.," and was for a time survived by his own father, the first Roderick, who had been born in 1753 and had fought at White Plains. The Beebe ancestry in America went back four more generations to one John Beebe, born in Broughton, England, November 4, 1628, who emigrated to Connecticut about 1650. His father, also named John, died aboard ship before reaching the colonies.[6]

William Beebe puts the matter succinctly when he describes his ancestors as having been "almost wholly of the good old British mixture of Viking, Anglo-Saxon, and Norman."[7] (The name itself appears to come from the Norman *de Boébé.*) From somewhere in this ancestry he inherited his tall, spare, well-muscled frame, his "six-foot self" which, at 160 pounds, he described as "indifferently upholstered." With this type of physique went a temperament showing lively mental and emotional perceptions and a high level of physical activity. As an adult Beebe asserted that he could get along on five or six hours of sleep throughout the year, and even as a boy he was notably active. He tells us of pole vaulting and kite flying and winter hiking and exploring and collecting; clearly he was not a bookish lad or a prize student, though he read a great deal and earned a solid B average in high school.

And somewhere in his early years was nurtured that extra

element of his heritage which foreshadowed his later distinction, that special conjoining of attributes which set William Beebe apart from hundreds of other boys in his home town of East Orange. That his boyhood was generally a happy one, in the usual sense of the term, appears altogether likely; his references to it are nostalgic and affectionate, if occasionally rather wry. He recalls *The Night Before Christmas,* the "child's beloved legend" of Santa Claus, and Christmas presents especially to be cherished. He remembers reveling in Barnum's circus even before he had learned to write, and many decades later he could vividly evoke "the smell, delectable and fearfully exciting in former years—of elephants at a circus."[8] He speaks of early travel to ancestral New England—where by a millpond he first identified the spotted sandpiper—and to Nova Scotia and New Brunswick and the Bay of Fundy. He tells us that on those magical nights in May and September when the migrant birds call across the dark skies, he had an urgent wish to turn a vast searchlight upward to see them as they passed.

But for a youngster so imaginative and emotionally aware, and so often alone in the necessary way of an only child, there were times of strangeness and fear as well as joy. So it was with kite flying, often enough the careless sport of breezy days in the sunshine—and yet also a source, as Beebe tells us, of "my boyhood's very real terror:"

> I wonder, if at some momentous happening in life everyone does not have the sudden recurrence of an emotion which has not been experienced since early childhood. Mere height or depth never affected me,—I could always look with pleasurable exhilaration over the edge of a precipice or down from a roof. But sometimes under the stars, when there came the realization of cosmic space, or at my first glimpse of moon mountains through a telescope or my first trip in an airplane, —then I shuddered to my soul, and my heart skipped a beat. I remember pulling in a kite with all my might, trembling with terror, for I had sensed the ghastly isolation of that bit of paper aloft in sheer space, and the tug of the string appalled me with the thought of being myself drawn up and up, away from the solid earth.[9]

To the "aching joys and dizzy raptures" assigned by Wordsworth to romantic childhood, Beebe adds a childish dread of isolation and abandonment, and the terror of nature's majesty and her unimaginable distances. Thus both joyfully and fearfully, and as much in the fashion of William Blake as of Wordsworth did

William Beebe apprehend the natural world, first as a child and later as a man.

Not long after his arrival in East Orange William had resumed his schooling at the Ashland Grammar School on Mulberry Street, and in September 1891 he entered East Orange High School. When he left in 1869 he had finished the major portion of his formal education, though perhaps he did not know it. At the high school he had taken four years of Latin and two of German—useful languages for the future naturalist, one for dealing with taxonomy, the other for reading the works of German scientists and explorers. He had had four years of English and one of English literature, and a semester of rhetoric; perhaps in the process he developed the keen language sense which would distinguish the twenty-odd books he was eventually to write. But the future Beebe is best foreshadowed in a series of six courses, each of one semester, which he may have taken as postgraduate work: physics, chemistry, geology, botany, physiology, and zoology. Here he achieved his best marks, ranging in the order given from the middle eighties to the nineties, and here he prefigured the career he was to follow.

These high school natural science courses provided only one outward sign among many. Young Beebe had begun his quest long before with creatures bearing such exotic names as Polyphemus, Promethea, and *Danais archippus* or, more prosaically, Monarch—moths and butterflies, to be pursued and captured, painstakingly if rather crookedly mounted in cigar boxes, and admired both as things of beauty and trophies of the chase. He had patiently hammered away at rocks in New Jersey quarries, perhaps to discover fossilized seashells or crinoid stems or even an occasional trilobite. He had begun the study of botany by learning the common wild flowers. He had read serious works of natural history by Jean Henri Fabre and Alfred Russel Wallace and Henry W. Bates, as well as "old tropic tales" that set him dreaming of faraway beaches curving white in the sunshine. But most especially, he had embarked upon the pursuit which would dominate much of his professional life, the devoted study of birds.

There are those people, Beebe wrote many years later, who seem to be born to take up birds as the prevailing interest of their lives. And he was such a one, finding in the crow's caw his best alarm clock, in the whistle of the chickadee a language he was delighted to learn, and in the books of John Burroughs, the famous "John o' Birds," nature stories to inspire his boyish emulation. In those days he was studying the bird life of a small park in East Orange, and he later recalled the sight of two Bohemian waxwings, birds quite rare this far to the east and south, as one of the great moments of his venture. Then from the nest of a house sparrow behind an attic window of the Beebe home he had purloined and blown the first egg of his collection, and when he was fourteen he began a collection of birds' nests, noting among other things that the house sparrow's untidy ball of twigs and grass was the only covered nest to be found among the various sparrow species.

At length he was permitted to own and use a gun, emblematic in those days of the serious bird student. About two and a half miles northwest of the Beebe home lay Orange Mountain, a forested ridge where the migratory species came in the spring and fall, and where the resident woodland birds could be found nesting. There William Beebe gathered the larger number of the specimens described by a colleague seventy years later as "a collection of well-made bird skins, mostly labeled 'Orange Mountain' and carefully sexed and identified"[10]—a collection essentially marking the beginning of his professional career. The writing career to which it was so closely linked began with a letter to the editor published in *Harper's Young People* for January 15, 1895, concerning the brown creeper.

Beebe's last years in East Orange saw him listed in the town directory as "Student," indicating somewhat ostentatiously his final semesters at the high school and his three years at Columbia University. Subsequently Beebe claimed a Bachelor of Science degree from the latter institution, but this is incorrect.* In fact he matriculated as a Special Student in the Department of Zoology in October of 1896, 1897, and 1898. In his first year he is on record as taking Biology II, in his second Zoology III, and his third Zoology 5. These three courses did not nearly suffice for graduation, nor did they support the claim made later that Beebe missed obtaining a degree only because he decided instead to take a job in 1899, at the beginning of his putative senior year. And if his college transcript fails to uphold him on either account, the same is true

* There is on record a somewhat testy letter, dated 30 October 1950, from one Halsey (Milton Halsey Thomas, Curator of Columbiana) to another university official flatly stating that Beebe never took a degree, despite his assertion to the contrary in *Who's Who*, and hinting at a further letter of preemptory inquiry to Beebe himself.

of possible collateral evidence. A search of alumni records and class yearbooks from 1895 through 1902 reveals no mention of William Beebe as a candidate for any degree or as a member of any class. On the strength of a high school diploma (even with some postgraduate credits) he could not have been engaged in a regular doctoral program, nor was he ever properly enrolled for the baccalaureate.[11]

But to deny Beebe his later academic claims is not to reject or depreciate the college experience itself. For a young man of intellectual zest and spirit, formal courses often do not provide the major rewards, which instead may spring from more informal arrangements and personal acquaintanceships. The special lecture series, attended not for course credit but for one's own enlightenment, was one such resource. In 1896, some months before Beebe first matriculated, the Academy of Medicine and the Biology Department had offered lectures under the general title of "Birds, Their Habits and Instincts," four of them delivered by Frank M. Chapman, ornithologist at the American Museum of Natural History, four others by Professor C. Lloyd Morgan, and one called "Ancient Life of North America" by Professor Henry Fairfield Osborn. A year later the American Museum sponsored lectures in the fields of botany, anthropology, and ethnology, including one on the American Indian by Franz Boas of the Columbia faculty. Then in 1899 came a series of ten lectures on the foundations of zoology (with particular emphasis on evolution) by William Keith Brooks, a visiting professor from Johns Hopkins University.

There were also courses which Beebe, as a free-ranging special student, might have audited to his own considerable benefit. Columbia had rather recently passed through a period of revitalizing change, with favorable consequences to the reorganized Faculty of Pure Science, which nearly trebled its staff between 1893 and 1903, and in the biological sciences offered courses in botany, biology, zoology, physiology, anatomy, pathology, and bacteriology.

If the record fails to show whether or not Beebe availed himself of these resources, it does indicate that he made personal contacts at Columbia of enduring value. Foremost among those he studied under, and second only to Theodore Roosevelt as a lifelong influence, was Henry Fairfield Osborn. Twenty years Beebe's senior, Osborn was both mentor and influential friend, and a man whose reputation in natural science was already well established when he came to Columbia in 1891 to reorganize the department of biology. He served for a time as Dean of the Faculty of Pure Science, and as department chairman in zoology until 1899. But increasingly his restless and dynamic spirit urged him toward organizational and administrative work with a strong public bent. He became president of the American Museum of Natural History—where Beebe, under the guidance of his mother, may have met him initially—and was named first president of the fledgling New York Zoological Society in 1899. Whether he prevailed upon Beebe to accept a job there or Beebe successfully urged his own candidacy is not the main question. The important fact, both for William Beebe and the society itself, was that a relationship was here inaugurated which ceased only with Beebe's death nearly sixty-three years later.

When Beebe took up his post of assistant curator of the bird house in mid-October 1899, the society's Zoological Park (later called the Bronx Zoo) was within a few weeks of opening its gates to the public. For a young man with considerable field experience but little formal training in ornithology and none at all in aviculture, it was not an easy time. His colleague Lee S. Crandall and his friend Fairfield Osborn, son of Beebe's mentor, recall those days as ones of trial and perplexity as well as achievement. Crandall notes especially the primitive state of understanding in America regarding the keeping of wild birds in captivity—then an "unnatural" practice in the eyes of many—and gives special credit to the Englishman Samuel Stacy, trained in aviculture at the Zoological Gardens in London, for the success Beebe had. Osborn emphasizes Beebe's plans and aspirations, in particular those regarding new and more spacious enclosures for exhibiting birds. Beebe disliked the cramped quarters of the first bird house, and saw many of his hopes realized when the Large Bird House was finished in 1905. And, according to Osborn, he proposed a flying cage as big as a football field, to be built outdoors. This too was eventually completed, but on something less than half the original Beebe scale.[12]

But purposes more strictly scientific than zoo display moved the Zoological Society also—in particular, technical research and publication, again activities largely foreign to Beebe's experience. Up to 1900 his published work (none of it of much consequence) had appeared in popular journals or in Frank Chapman's nontechnical magazine *Bird-Lore*. Beebe continued this practice in the early years of his curatorship by publishing many articles in the semi-popular *Zoological Society Bulletin*, in *Recreation* and *Outing*, and even in the *New York Tribune* and the *New York Evening Post*; and indeed, he never gave up writing for the popular audience. However, with more serious endeavors now expected of him, he undertook various researches and appeared for the first time in *Science* and *The Auk*, the latter a journal published by the American Ornithologists' Union. And when in 1907 the Zoological Society brought out its own technical journal *Zoologica*, Beebe's long article on "Geographical Variation in Birds with Especial Reference to the Effects of Humidity" dominated the first issue. This article was the result of a series of controlled experiments Beebe had made, with the intriguing result of showing that artificially induced humidity produced darker and darker plumage in succeeding generations of the subject species. Here he had made an auspicous beginning in technical writing for *Zoologica*, where a hundred other Beebe articles in many areas of natural history were to appear in the decades to follow.

By this time Beebe had also published his first technical book, *The Bird, Its Form and Function* (1906).[13] Given the fact that some of the chapters had already appeared in *Outing*, *Bird-Lore*, and the *New York Evening Post*, the word "technical" need not be taken too literally. Yet among the seventeen chapters are those dealing with paleontological evidence of bird ancestry and evolution and with avian anatomy, physiology, locomotion, nidification, and embryological development. Obviously Beebe had done the appropriate research among the proper authorities—a fact which no doubt gave pleasure to Henry Fairfield Osborn, to whom the book was dedicated "by His Former Pupil, the Author."

In the Beebe canon this book is something of an oddity. It seems very much a first book, a piece of apprentice work, though by date of publication between hard covers it is not. Its tone is ambivalent, its voice uncertain, suggesting an author divided between cool data and warm responses, the objective and the interpretive. Hence the youthfully exuberant flights of rhetoric in unlikely places, the common usages employed amid arcane materials, the straining for effects that do not come off. The mixture is clearly not a stable one. An author in such a state must decide whether he wishes to write books about his scientific adventures, or books of science. *The Bird* was ostensibly one of the latter—but there would not be another one quite like it.

Later in 1906 Beebe published another book, *The Log of the Sun.*[14] Even more than *The Bird,* this was a collection of pieces which had appeared earlier in the popular prints, newspapers included; and if *The Bird* was intended as an introduction to ornithological science, *The Log of the Sun* proposed "to inspire enthusiasm in those whose eyes are just opening to the wild beauties of God's out-of-doors." The former book is dedicated by a pupil to his professor, the latter by a son to his parents, "whose encouragement and sympathy gave impetus and purpose to a boy's love of nature." What follows is a set of chapters making up "A Chronicle of Nature's Year"—such at least is the claim. The range is very wide, taking in virtually all of the life sciences and extending to meteorology, geology, and paleontology. The principal idiom is the essay, often rather brief and occasionally didactic, with quotations from various sources that range in quality from Thoreau's sinewy *Journal* to W. H. Carruth's flaccid verse ending "Some call it evolution, / And others call it God." Again the reader has a sense of apprentice work—but with welcome changes for the better. There is little here of the contrived diction of *The Bird,* the strained brightness, and the importunate familiarity. Some of the essays impart rather obvious lessons, others edge toward anthropomorphism or religiosity; characteristically, however, Beebe expounds lucidly and describes aptly, and offers comments only in a genial and unobtrusive way.

The older nature writer who comes first to mind is John Burroughs, then at the height of his fame. In him Beebe had a worthy model, sedulously to be aped; and yet even this early the Beebe individuality asserted itself. John Burroughs was generally rather sedate, whereas in *The Log of the Sun* we already discover some of the high spirits and even a touch of the astringency which would mark Beebe's later work. And by this time also Beebe was exhibiting a range of scientific knowledge which Burroughs never commanded, and an abiding curiosity which Burroughs shared but seldom equaled. Which is simply to say that William Beebe was fast becoming his own man, and the contest, if such it was, between books of science and those of scientific adventure was being decided in the latter's favor.

Chronologically, in fact, the decision had already been foretold. First of his books in respect to date of publication, if not necessarily of composition, was the adventure book *Two Bird-Lovers in Mexico* (1905).[15] In 1902 William Beebe had married and soon had set out with his bride on the first of several journeys which would serve as sources for books, not merely of recorded fact but of personal quest and discovery.

:2 Journeys with Mary

> It was an act of fate that at what is considered a most impressionable age (though as a matter of fact what age is not impressionable?) I happened to go to Mexico . . . Mexico was my introduction to Latin lands. Yet it all seemed oddly familiar, as though somehow I had known it before. I felt quite at home; with that sense of belonging which comes whenever crossing the Potomac bridge I find myself in my native Virginia.
>
> *Journeys in Time*

The better part of a lifetime had passed between the Mexican journey recalled in these lines and the mention so belatedly made of it.[1] And an earlier "act of fate" is not spoken of at all—the marriage of a girl named Mary Blair Rice to William Beebe at Clarkton, Virginia, on August 6, 1902. That winter the young couple had gone to Florida and "had reveled among the angel fish, the corals and the sponges" of the Keys. Then late in 1903 they set out on the Mexican venture which, forty-three years later, still seemed fateful to the woman who had been William Beebe's bride.

That she should speak lightly of affinities between Mexico and her native state was most unlikely. Her Virginia ancestry, like Beebe's in New England, went far back, tracing its beginnings to Theodorick Bland, who came to the colony about 1654 and settled at Westover on the James River, a site later made famous by William Byrd. It was at the village of Coles Ferry on the James that Mary had been born to Henry Crenshaw Rice and Gordon (Pryor) Rice, daughter of General Roger A. Pryor, and it was on the family plantation nearby that she had been raised and largely educated. With the nearest railroad twenty miles away, such a plantation was necessarily "culturally, as well as economically and socially, self-contained"—but little cultural deprivation is suggested by the authors Mary found to read in the family library: "Grimm and Andersen, Lewis Carroll, the *Arabian Nights;* Scott, Dickens, Thackeray, Jane Austen, George Eliot, Anthony Trollope, the Brontes; Poe, Emerson, Thoreau; Shakespeare, Flaubert, Victor Hugo; Tolstoi; and, later, Kipling and Conrad; with some assorted poetry."[2] If, as one biographical source states, Mary Blair Rice was schooled for a time in Massachusetts, some notably sturdy foundations of learning had already been laid. And though she was far from unlettered when she married Beebe, she continued to develop creatively and intellectually long afterward. As Blair Niles she eventually became a highly regarded writer, but as Mary Blair Beebe she first showed her considerable talents.

The Mexican trip began with the departure of the Beebes by ship from New York on December 17, 1903. This was to be the first major bird expedition undertaken since Beebe's appointment to his post. Earlier bird trips, to Nova Scotia, Cape Cod, Virginia, and Florida, had been measured in days or weeks; this time it would be four months. The ship touched briefly at Havana and on December 24 approached the coast of Mexico. "The sun sank into a sea smooth as glass, and when its golden path had faded out, a tiny thread of silver was left,—the thin moon-crescent hung even-balanced in the western sky,—and our last night on the water—our first Christmas Eve in the tropics—was one of enchantment."[3] They landed Christmas Day at Vera Cruz and arrived January 1 at Guadalajara, Jalisco.

Obviously Beebe's primary task was to discover and identify and collect as many Mexican birds as possible. This was a new avifauna to him, and the areas chosen for study, mainly in the states of Jalisco and Colima, were far enough south of the U.S. border that species common to the two nations did not predominate. Nevertheless for much field work he seemed to depend on works in English, in particular Florence Merriam Bailey's *Handbook of Birds of the Western United States* (1902). He also chose a variety of habitats in order to encounter birds identified with each such area. Thus the trip was divided into periods of about two weeks each: early January around Guadalajara, late January in the higher barrancas (ravines) near the eastern slopes of the volcano of Colima; early February in the lower barrancas of the southern slopes, late February in the hot lowlands west of the volcano and the coastal region not far from Manzanillo; then back to Guadalajara in March, with a trip to Chapala and its marshes concluding the month. So the Beebes ranged hundreds of miles overland and from a mountainside at 4,000 feet to the shores of the Pacific Ocean.

They traveled much on horseback and lived mostly in tents. Both carried revolvers, not an unusual practice in a country still politically unsettled, despite—or perhaps because of—the harsh dictatorship of Porfirio Diaz. Such traveling and camping as they did, with a few pack animals and hired help temporarily acquired, was scarcely a well-established custom; in fact the Beebes found no one at all to advise them on Mexican camping. By no means dismayed, they set forth anyway, for they were both of an age to be not merely impressionable, as Mary said, but happily heedless.

Their spirits were well rewarded. Here is William Beebe speaking of an adobe dwelling they shared: "How close to Nature one seems to live thus! closer to Mother Earth than did Thoreau at Walden; and yet when this framework of mud is clothed within with clean plaster, in rooms cool-tiled and with ceilings of taut linen, sleep and study and the joys of very life come in pleasantest forms." And of their first night in camp by a stream near the base of Mount Colima: "The sleep that came to us that night was of the quality of sweetness known only to those whose happiest days and nights are the ones spent closest to the heart of wild Nature." Then from Mary Beebe: "Though I were to write a volume I could not adequately picture the great charm of our wild free life in camp! One lives so near the heart of Nature, and in this simple natural life learns many a great truth. . . . What a glorious thing is a cold plunge in early morning in the swift-flowing river near the tent, where the night before the deer drank, and along which all the furtive wild creatures of the night stealthily made their way in the moonlight. Here one feels how good a thing it is to be alive, to be hungry and to eat, to be weary and to sleep."

Though Mary wrote the last chapter for *Two Bird-Lovers in Mexico,* William did the rest of this relatively long book. Here he was not compiling chapters written over many months on various subjects, as in the case of *The Bird* or *The Log of the Sun;* basically he was telling the story of a three-month search for birds in certain regions of one country. That he was writing of a second honeymoon is fairly plain also, with delight in his recent bride as well as his new birds. But much of this is implicit; the birds and the journeys to find them predominate. Except for the final chapter, and despite an occasional fond reference and the dedication of the book "To My Wife, The *Other* Bird-Lover," Mary Beebe remains a background figure. Her reactions are seldom given, she is almost never quoted, she is not shown at her own work or making her own decisions. She exists mainly as the unheard other half of the plural pronoun, and when she is referred to explicitly it is as "the Señorita" or "My Lady" and not as Mary.

Simply as a piece of writing, *Two Bird-Lovers in Mexico* might have been improved had the sexual theme been allowed to provide something of a unifying principle, as it does in a special way in W. H. Hudson's *Green Mansions.* Of course Beebe ostensibly was writing of an ornithological expedition and not of romance—but in fact he was doing both, unless one may dismiss his (and Mary's) apostrophes to physical and psychic joys as pleasant verbiage coolly interpolated to beguile the reader. Neither personality accords with such a notion. The erotic was central to them both, but in the book as written, it remains quite peripheral to the birds, the geography and the climate, the trails and pack trains and native helpers, even the towns and cities. Beebe was to learn how to select and shape such diverse materials with increasing art in later books, and to deal with the erotic in a quite special way in *Pheasant Jungles,* his most ambitious popular work. That time was not yet, and this Mexican book, for all the power of its separate passages or chapters ("The Magic Pools," for example) remains a prisoner of its data, a kind of extended bird walk in an interesting foreign land.

Mary Beebe's chapter "How We Did It" seems on first reading to be rather an artless addendum, full of such earnest advice as *"ignore small discomforts"* and such homely lists as those for apparel and cooking gear. A closer reading suggests strong individuality and a spirit of engaging zest and hardihood. Ammonia should be taken along, Mary says levelly, to treat bites of "great hairy-legged spiders" and the stings of scorpions. Sidesaddle riding is quite out of the question over this rugged country, hence a divided skirt must be worn, though "Mexicans are much surprised at seeing a Señorita ride cross-saddle and wear a revolver." Veils or a net for the hair are useful when riding, but "I hope my camping woman will not mar her pleasure by wearing her veil *over* her face. A wild gallop over the plains loses much of its charm if there is anything between one's face and the pure invigorating mountain breezes. And after all, a little honest tan is a good thing!"

There was a great deal of riding to be done on this Mexican journey, first in the area around Guadalajara, later to campsites around Mount Colima, then an early morning gallop of many miles along a jungle trail on the way to Manzanillo, and finally a sixty-mile ride (ending at four in the morning) to catch the train for the return journey. The reader does not doubt that these were fit challenges for a seasoned horsewoman; then he learns the truth in almost the last paragraph: "As to horseback, my theory is that all one has to do is to get on and *ride.* I have little patience with spending months learning to ride. I had never ridden before, but I simply got on and rode off. Of course for the first few times one cannot ride long distances, but that soon comes with a little practice. The rule for a good dancer applies equally to a good rider—do not be rigid, let yourself go."

In their next book, published in 1910, the Beebes shared almost evenly in the writing tasks and fully in the title page, which reads in part: "*Our Search for a Wilderness* by Mary Blair Beebe and C. William Beebe, Curator of Ornithology in the New York Zoological Park."[4] Mary wrote the third chapter, William the ninth, tenth, and eleventh; all others were jointly composed. Note should also be taken of the dedication: "To Judge and Mrs. Roger A. Pryor With the deepest affection and admiration of their Granddaughter Mary Blair Beebe and of C. William Beebe." Here were two of Mary's more remarkable progenitors, offering proof if need be of her claim to membership in the First Families of Virginia. Roger Atkinson Pryor (1828–1919) had been a member of Congress in 1860, had served as a Brigadier General in the Confederate Army, and finally had become a judge of the Supreme Court in New York City. And Mrs. Pryor—Sara Agnes Rice Pryor—had already published four books, including *Reminiscences of Peace and War,* brought out by Macmillan in 1904.

The "search" referred to in the Beebe book was a series of trips taken over a two-year period, and made possible by a leave granted by the Zoological Society. There was, however, no grant of money; these trips, like all the previous ones, Beebe financed privately. The explicit meaning of the title comes out in this passage: "Was there no spot left on earth, we wondered, which could truthfully be called an untrodden wilderness!—jungles untamed by axe or fire, where guns had not replaced bows and arrows; where the creatures of the wilderness were tame through unfamiliarity with human beings!" The answer was somewhat equivocal: if by "untrodden" they meant undiscovered by whites, then their search failed; but if they meant unfrequented or little known, they found much that fulfilled their hopes.

On the first journey the Beebes left New York on February 22, 1908 on the Royal Mail Steamboat *Trent,* and after touching at several other ports, arrived at Port of Spain, Trinidad, on March 9. On the evening of March 24 they left Port of Spain and crossed the Gulf of Paria on the 21-ton sloop *Josefa Jacinta,* reaching the outlet of the Rio San Juan, west of the main Orinoco delta, after a night and day of heavy weather. Ascending to the little village of La Ceiba on the tidal river called Guarapiche, they explored this strange water-world of mangrove forests for several days, making the anchored sloop their home. Then by another tidal passage they reached Guanoco, the small port serving the pitch lake called La Brea—the name evidently given by Spaniards wherever they discover such a place, whether in southern California or southern Trinidad or, as here, northeastern Venezuela. At La Brea the Beebes stayed at the headquarters of the company engaged in extracting the pitch, and were served by the staff. In the area roundabout they had excellent opportunities to study a variety of wild creatures and to collect live birds for the New York Zoological Park—in all, forty individuals of fourteen species. After returning to the gulf by the San Juan, they left Venezuela at midnight of April 14 "on a large tug used by the Pitch Lake Company" for Port of Spain and eventually by steamship for New York.

The second trip was divided into three parts, each in a different area of British Guiana (present-day Guyana). Having arrived by steamer from New York at the Guianese capital of Georgetown on February 24, 1909, the Beebes left a week later by sea for Morawhanna, near the Venezuelan border. While awaiting a launch to take them to Hoorie Mine, they took a side trip to Mount Everard. The trip to the mine was three days by tent boat up the Waini and Barama rivers to Hoorie Creek. After a week they returned by Indian canoe through a complex and isolated series of waterways to the Essequibo estuary and thence to Georgetown, reaching there on March 21.

Two days later they started again for the Essequibo by small steamer, changing at Bartica to a launch and then to a tent boat as they proceeded up the Mazaruni and Cuyuni rivers to the Aremu and Little Aremu, stopping at the Aremu Gold Mine. After some weeks they made the return journey by the same route, but running many of the rapids instead of portaging around them.

The final excursion was much the shortest, both in time and distance. The Beebes left Georgetown early on April 12, taking the train to Abary Bridge two hours east of the capital, and going by launch three hours up the Abary River to a bungalow on a small island by a lagoon just off the river. They planned to spend a week, but on the morning of April 15, Mary broke her wrist in a fall and they quickly returned by launch to the rail bridge and thence to Georgetown by an emergency train summoned by a message sent ahead. They sailed from Georgetown for home on April 24. From their excursions in Guiana (with the assistance of Lee S. Crandall) they had assembled 280 live birds of fifty-one species, thirty-three of the latter new to the Zoological Park collection. And of course they had gathered a wide range of data both for technical work and for their next book.

Perhaps again, for purposes of a book, the range was too wide. Although each of their four excursions had taken place in the northern coastal regions of South America, together these trips had encompassed a variety of habitats and provided diverse experiences. For all their zest and involvement, the Beebes in *Our Search for a Wilderness* fail to select with sufficient rigor among their materials, especially in writing passages of extended sequential action. Perhaps in a way they were being too honest, relating events more as they actually befell than as art requires. In any case, as with the Mexican book, details tend to clutter too many of the pages, detracting by their very number from passages of vivid natural beauty or high scientific adventure.

Joint authorship seems to impart ease to the handling of personal relationships, and to bring Mary Beebe more clearly into focus. But as before, Mary writing alone is her own best portraitist. Her chapter deals only with the Venezuelan journey of 1908, and offers new material only in that she retells much of the story from her own point of view. Thus the first night of the voyage across the Gulf of Paria—already described in fairly cheerless terms—becomes a sort of contest between the legion of rats that raced and fought and tumbled and squeaked just beneath her "catacomb" bunk on board the *Josefa Jacinta,* and her determination to get some sleep. Finally she decides to sleep on the floor, without notable success. She then offers comments on the ship's cook, a man so formidably unsanitary that she ruefully remarks: "We so often wished we had brought graham flour. White flour does show the dirt so!" And as to his custom of hanging fresh salted beef to cure on the sloop's rails and rigging, she admits her dismay at the attendant stench, but bravely reflects, "suppose it were fish."

At the pitch lake headquarters there is a good deal of talk among the Venezuelan members of the staff about political matters, and in the midst of one boastful colloquy she smiles at a particularly vainglorious assertion before realizing that she is giving offense—and she frankly records a gentlewoman's mortification that she has done so. Later she tells a charmingly forlorn story about an even more trying social situation. Having volunteered to gamble at "siete y media" with three men at the headquarters, she finds herself losing steadily all the way to midnight, "woefully tired and sleepy" but unaware that the loser has the right to call the game whenever he (or she) chooses. William, who has long since gone to bed, so informs her in the morning, but with what seems to her a cruel lack of sympathy.

Surely there is little of the Pollyanna in Mary Beebe, and even less of the complainer. She forthrightly states, for example, that life aboard the sloop was plagued by night rains and mosquitoes and unlovely odors and various noises from the crew; yet she writes: " 'How *could* you enjoy it?' I am often asked: but the trifling discomforts were all in a day's work and more than compensated by the beauty and freedom and wonder of it all. They served to make us know that it was not all a dream." And in another passage: "Never were nights more beautiful than those we spent on the deck of that little sloop, and never was sleep more dreamless and peaceful. . . . W——— and I lay, as we so often did, staring wonderingly out into the night,—the marvellous tropical night."

William for his part responds somewhat in kind, especially in dealing with Mary's accident in a chapter he wrote alone. There is more in his account than an affectionate husband's concern; there is anxiety that borders on fright, and fond solicitude that is almost uxorious. And he too has a rhapsodic passage to sum up the experiences they have had together: "[As we depart] we know in our hearts that someday we shall return. Meanwhile the thought of that vast continent, as yet almost untouched by real scientific research; the supreme joy of learning, of discovering, of adding our tiny facts to the foundation of the everlasting *why* of the universe: all this makes life for us—Milady and me—one never-ending delight."

Before the year was out they had set forth on the greatest expedition either of them would ever take, a quest for the pheasants of the world. Decades later Mary wrote again of the "fate" which took her "to England; to what we absurdly call 'the Continent,' as though there were no other; to Egypt, Ceylon, India, Sikkim, Burma, Yunnan, the Malay States, Java, Borneo, China and Japan." In his eight-page report to the Zoological Society, William began and ended the account of their journey thus: "Mrs. Beebe and the writer left New York for London on December 26, 1909. . . . We reached New York, completing the circuit of the globe, on May 27, 1911. Altogether, Mrs. Beebe and myself spent seventeen months in this search for pheasants, visiting twenty countries and travelling approximately fifty-two thousand miles."[5]

An undertaking as formidable as this, involving not only the cost of travel itself but the outfitting of many separate journeys within the overall expedition, the hiring of servants and guides and porters, the packing and shipping of specimens, and later the assembling of the scientific and artistic results into the most imposing publication of Beebe's career, was scarcely one to be financed from an ornithologist's private purse or from the limited funds of the Zoological Society. But Beebe had found a donor both affluent and enthusiastic about the pheasant project, Colonel Anthony R. Kuser, a public utilities executive, financier, and philanthropist with a special interest in raising pheasants on his estate at Bernardsville, New Jersey. Nor could such a task be accomplished as part of regular duties at the Zoological Society; Beebe took a five-year leave of absence, 1910 through 1914, to carry out his field work, his museum and laboratory and library research, and his writing and editing for the four great volumes published as *A Monograph of the Pheasants* between 1918 and 1922.[6]

Only very tamely does the term "field work" suggest the dimensions of Beebe's enterprise or the vast reaches of the eastern world that he and Mary would explore. Among the twenty-odd genera of pheasants and their allies which Beebe resolved to find and study are some of the most resplendent birds in the world—and some of the most inaccessible. They can and do live in the equatorial jungles of Malaya, in the Himalayas all the way to snow line, and in the remote desert uplands of Mongolia. They include the peacock, known and celebrated since ancient times, the jungle fowl, ancestor to the domestic chicken, and the ring-necked pheasant, widely and successfully introduced in regions far from its native Asia. But they also include Sclater's impeyan, the bronze-tailed peacock pheasant, and the ocellated argus, never seen alive in the wild state by white men until Beebe set out to do so—and the argus, though found dead in his traps, still eluded him.

The *Lusitania* took the Beebes across the Atlantic to Britain, but for most of the journey their vessels were far less imposing. From Brindisi, reached by train from France later in January, they took the British mail boat *Isis* through two days of violent storm to Port Said—with William, at least, seasick all the way. R. Bruce Horsfall, a well-known bird artist, joined them there, to remain with the expedition for six months. Another ship took them to Colombo, Ceylon, and then the small coastal steamer *Lady McCallum* ferried them to the village of Hambantotta on the island's south coast. From an anchorage offshore the Beebes and their full array of equipment were landed by outrigger canoe. Then in conveyances equally humble, big lurching bullock carts, they were taken over ten miles of rough trail to the jungle community of Welligatta, thus beginning in a manner both symbolic and joltingly real the actual quest for the great birds. There William found the jungle fowl and saw for the first time in his life a peacock in the wild. Subsequently they reached the center of the island, where Mary, as she recalled much later, "watched Mahouts bathing their elephants in the river at Kandy."

The Beebes then sailed from Ceylon to Calcutta and early in April proceeded by train northward to Darjeeling. "With thirty-two Tibetan men and women coolies we left this last outpost of civilization and on small Tibetan ponies, made our way northward over difficult trails and through the most magnificent scenery in the world. With Everest and Kinchinjunga in full view we pushed on higher and higher until we passed through every zone up to the very snows." So William reported this journey in 1911; and Mary remembered how she had ridden "day after day through the unearthly loveliness of the Himalayas." With such beauty, however, came certain penalties—not only snow and biting cold in the higher reaches, but air so rarified that any major exertion quickly tired them. But by taking a number of excursions in this region where Nepal, Sikkim, and Tibet come together, they "ultimately found and studied, at various altitudes, all the groups of eastern Himalayan pheasants."

Early in May the Beebes returned to Calcutta and almost at once set out for the Himalayas a thousand miles to the north and west, the mighty Hills that crowd the border between Garhwal and Kashmir and Tibet. Here among forests of deodars and spruce clothing the great ravines Beebe sought out and observed and collected the koklass and cheer pheasants, the great tragopan, and the bird he thought of as the most beautiful of all the pheasants, the impeyan. He also had the satisfaction of discovering that the tragopan makes its nest in a tree, a fact never before verified.

With the hot season already closing in on the Ganges plain, and the rains that inhibit good field work coming to Assam and Burma, the Beebes next took ship from Calcutta for Singapore, almost exactly on the equator, to set up a base of operations for a summer of work.* Their first excursion was eastward to the British enclave of Borneo called Sarawak. Here at Fort Kapit they hired a seventy-foot Dyak war canoe and a dozen paddlers to take them up the Baleh and Mujong rivers to the haunts of five Bornean species of pheasant, including the white-tailed wattled, the sempidan, the fire-backed, and the great ruoi or argus. In this jungled river valley they lived for weeks as two white people alone among native headhunters—and found them to be hardy and fearless, and also tolerant of odd western ways and willing to contribute their jungle lore to this strange scientific enterprise.

Their excursion to Java took the Beebes the entire length of the main island and then to the island of Madura just to the north. They also visited Belitung lying between Borneo and Sumatra. Then a relatively short trip north from Singapore to Kuala Lumpur began their extended and often arduous expedition in Malaya. First they made their way by train and mail truck to a government bungalow at Samangko Pass, 3,000 feet up in the central mountain range, where tree ferns grew in thick forests and the bronze-tailed peacock pheasant and the great ocellated argus challenged Beebe's skills. His success with the first was counterbalanced by his failure with the second. On more than one night Beebe lay listening and waiting in his hammock or crept through the dark far into the jungle, following the argus to catch a glimpse of the living bird by

* Disagreements with Beebe at this point prompted Horsfall to leave Singapore for home.

flashing his powerful light at the right moment—but always he failed.

Stopping for side trips along the way, the Beebes at length reached Kuala Lipis, a stifling town almost in the center of the country—and now threatened by a cholera epidemic moving up the Pahang River. After a trying week and an attack of fever that laid William low for a day or two, they engaged a government houseboat and "a crew of five Malays and a Chinaman" for a slow cruise on the Pahang and its tributaries, with many stops along the way for scientific forays in the jungles and swamps of this torrid lowland. Almost any walk through the jungle was impeded painfully by thorn-laden undergrowth, and unpleasantly by "myriads of land leeches, scores of which fed on our blood whenever we left the boat." Mary's brief mention of this period says nothing of such things, but years later a photograph appeared in the *Zoological Society Bulletin* showing a leech on the hand of someone in Malaya —and, on the third finger, a Tiffany-set engagement ring and a wedding band.

To escape the cholera zone along the coast, the Beebes at length left the boat and proceeded overland to a new rail spur over which they made their way to the main tracks and thence to Johore, just north of Singapore. Two shorter trips in Malaya concluded this equatorial phase of their expedition; and now that autumn had come and with it the decline of the rainy season in Burma, the Beebes sailed from Singapore late in October for Rangoon.

For the 300 miles by rail from Rangoon to Mandalay they traveled with tourists guided by travel agencies (or perhaps by Kipling's injunction) but for the remaining 350 miles to Myitkyina they were spared such trials. Myitkyina was scarcely a town to attract tourism; when the Beebes arrived they found a British punitive force about to set out toward the troubled Chinese border, and later, when they had proceeded some distance northeastward with their entourage of a personal servant, a cook, three mule drivers, fourteen mules, and three riding horses, they were required to add an armed escort of six Gurkhas before venturing further.

In this border region of steep mountain ranges, wild river gorges, and turbulent politics they spent the better part of two months, finding the white-eared, Lady Amherst, and silver pheasants, and naturally hybridized strains of the kaleege. It was here also that William became the first white man to see Sclater's impeyan alive in the wild. On an excursion into Yunnan, the westernmost province of this part of China, he also found portions of pheasant plumage which, when later correlated with specimens in a Paris museum, formed the basis for the new species *Ithaginis kuseri,* named in honor of Beebe's patron.

Toward the end of December they returned to Singapore, concluding their work in this part of the Orient by packing up their cases of specimens and shipping them home. On the last day of the year 1910 they set sail for Shanghai. From their subsequent experiences in China Mary recalled "the crooked narrow streets of such cities as Foochow, with their bright painted signs," and the feeling of standing at dusk upon the Great Wall; and William reported: "In Eastern China our plans were continually upset by unforeseen events such as sudden riots, terrific snow and wind storms, and the prevalence of the plague. . . . But by constantly re-adapting our plans to the new conditions we were able at last to reach the objects of our search; whether by steamer and sampan, as in the valley of the Yangtze; by house-boat, as in the region back of Foochow; or by palanquin and camel in the bleak deserts of Mongolia."

Japan, seemingly so placid and well-ordered, offered no such strenuous challenges. There the Beebes conducted the last of their studies together, sailing in the early spring of 1911 for home. The following year was their last as husband and wife. Early in January, 1913, Mary left William to return to her parents, and by February she was in Reno, Nevada, to begin the six months of residence then required for a divorce. The decree was granted in the court of Judge Cole L. Harwood on August 29, on grounds of "cruel and abusive treatment."

"NATURALIST WAS CRUEL" was the headline of a short news item next day on the front page of the *New York Times.* Mary Beebe was described as wearing dark glasses on the witness stand, thereby giving emphasis to her charge that constant typing and other close work for her husband had seriously impaired her eyesight. She further charged cruelty, indifference, even a threat of suicide: "Once, she said, he put a revolver in his mouth and threatened to shoot himself in an effort to frighten her. For days at a time, said Mrs. Beebe, her husband refused to speak to her. . . . There was no opposition to the suit today."

The melancholy particulars of her charges had appeared August 12 in a complaint of six legal-size pages.[7] No doubt unwilling to undergo the ignominy of a New York divorce (perhaps involving spurious evidence of adultery "discovered" by prearrangement) Mary had chosen instead the less demeaning alternative of Reno. But the bleak imperatives of the legal process required that in order to substantiate her allegations of cruelty and abuse, she dredge up details of eleven years of marriage, offered not in the context of early happiness and hope, or even of later doubt and frustration, but as discrete data, stripped of human complexity. So we read of financial stringency combined with reckless spending, domestic drudgery, unexpected guests, uncongenial neighbors; of verbal abuse, suicide threats, long silences, unexplained absences, public scenes, and at last insupportable discord and needful flight. Beebe's answer—a single page presented by his attorney, and involving no testimony on his part—denied all major charges but offered no details whatever in refutation. Mary's prayer "that the bonds of matrimony heretofore existing between plaintiff and defendant be dissolved" he opposed by praying "that the prayer of the plaintiff herein be denied." But the fact that he did not actively contest the divorce indicates that he was willing to see it granted. While Beebe's absence from the courtroom spared both parties the distress of personal confrontation, it also excluded from the record any testimony which might have countered Mary's charges, or at least suggested the complexities of a marriage relationship here seen only from the plaintiff's point of view.

One of the most interesting aspects of Mary's complaint is that its details are almost entirely domestic, in two senses of that word. Virtually all the alleged cruelties and indignities occurred within the household in New York, and not a single charge refers to events in Mexico or Venezuela or Guiana or anywhere in the Far East. A clear inference is that the many months the Beebes spent together on expeditions were happy ones, while everyday life in the city was a trial almost from the start. But further analysis suggests that discord in fact notably increased in the period from 1909 through 1912—seventeen months of which were taken up in search for pheasants. According to Mary, domestic quarrels during these years saw Beebe threatening suicide "in numerous ways, such as throwing himself in the river, shooting himself through the roof of the mouth with a revolver, and by cutting his throat with a razor." So it appears that discord was cumulative, and it is doubtful that the expeditions were wholly spared the tensions that grew over the years in New York. Perhaps the Mexican trip, almost a second honeymoon, was in truth the joyous venture they both described; later ones were doubtless otherwise, and the last occurred in the time of final alienation and parting.

Given the exotic character of Mexico—the first foreign land they visited together—and the novelty of their first real expedition, the Beebes may well have enjoyed the land and each other as fully as *Two Bird-Lovers in Mexico* suggests. Four years would pass, however, before their next major journey; and by Mary's testimony at least, these were years of struggle and disenchantment. *Our Search for a Wilderness* reveals little or nothing of this; but in William's protestations of supreme joy and unending delight and Mary's apostrophes to the shared wonders of tropical nights there was as much of brave hope as of reality observed in the sixth and seventh years of their marriage. Presumably the eighth and ninth years, those that embraced the long and frequently trying excursions in search of pheasants, offered other counsel.

In a manner oblique and yet unsparing, Mary herself suggests as much. Her introduction to *Journeys in Time* contains more than travel notes about the Orient; it also offers, as she says, passages which are "as near to being my own autobiography as anything I could ever write." Ostensibly she is discussing the "sense of belonging" she experienced when she and William first visited Latin America; but then she offers a suggestive contrast in her response to the journey to the Far East, again in William's company. Here she found "a dream which was not for me. I dreamed the dream, but I awoke knowing that I must return to the Americas where I belonged, and wondering at my instinctive identification with the Latin-American countries." Early love, the good years of Mexico and perhaps Venezuela, remain; what came later was the unhappy awakening from dreams no longer sweet.

Marital alienation is commonly the result of joint enterprise —and there is good reason to believe that William contributed in full measure to the process, not merely during the last years in New York but during the pheasant journey itself. Nothing in his formal report to the Zoological Society speaks to this point; but sixteen years later, when he chose at last to publish his pheasant tales in a book for the general public, he revealed, for whatever purpose, personal things odd and even unsettling. By his account, more than once during the seventeen months of his journey with Mary he suffered acute stages of agitation bordering on nervous collapse; and he states that once at least he shot a man dead for supposed aggressive acts. That the man was a native is clear; that he was shooting arrows—poisoned ones, Beebe charges—at the expedition camp is alleged; that Mary was present at the time her husband asserts that he killed the man with a rifle bullet is beyond reasonable doubt.

Not that William says as much; by this time Mary has vanished from his public work. Or nearly so; the persistence of Mary's shadow in his prose is intriguing to pursue. Some references are obvious and a bit snide, as when Beebe speaks of humble creatures as First Families of Vertebrates or First Families of Nonsuch. Others are self-serving or self-centered, as in the pretense that the perils of the pheasant expedition were confronted alone by a solitary and intrepid American male—or as in the statement that this same lone male observed and wrote up the events given in *Our Search for a Wilderness.* But mixed with such claims there may also be a note of wistfulness: "Seven years ago I passed this way en route from Morawhanna, paddled by six Indians. *Maintenant ce n'est qu'une mémoire.*" And regardless of the scientific reference, the experiences he shared with Mary, especially in the Orient, pervaded his life as no others before or after. He came to realize that those were days of greatest joy, "wonder times [that] lived only through memory and were misted with intervening years." And twenty-two years after his marriage, and eleven years after his divorce, William Beebe could evoke the past with Mary in this remarkable passage, occasioned by his departure from the Galapagos Islands:

> Daphne Major slipped by, the very ghost of an islet, a mere blotter-out of stars beyond. . . . So had Orizaba once faded from view, and so the jungle of Borneo and the lights of Rangoon; so had been erased the last silhouette of Kinchinjunga and of Fuji. Tonight, with the passing of Indefatigable, there came the faint aromatic scent of Bursera leaves; but whether this, or the perfumed breeze which blows from the camphor groves of Kagoshima, or exciting odours from the Calcutta bazaars, or the scent of white jasmine in a Virginia garden, such memories are eternal, they are the saddest things in the world, and they pass all understanding.[8]

A true heart cry, bravely put—and yet properly Romantic and thoroughly self-aware, as befits the man who chose the words and put them down. "Maid of Athens, ere we part, / Give, oh give me back my heart!" Byron knew that the dream must fade, the parting come; William Beebe knew that "never-ending delight" must end, for melancholy victory lies in fated separation, the heart borne down with rue and yet lifted by sweet memories of first love won and lost. And, like Byron, he would seek and seek again.

:3 The Great War

> Mr. Beebe is out here; he has just come from France; on the French front he was allowed to do some flying and bombing—not fighting the German war-planes.
>
> THEODORE ROOSEVELT
> *March 17, 1918*

Had William Beebe himself not chosen to emphasize his Great War experiences, others could pass over them lightly or simply ignore them. But no reader of Beebe can miss so recurrent a theme, or doubt its importance. Scarcely a book of his, from *Jungle Peace* in 1918 to *High Jungle* in 1949, is without its war references.[1] Yet he is a perspicacious reader indeed who can readily derive from all this any clear pattern, whether of actual events or Beebe's response to them. All that is clear is that the events greatly troubled his spirit, and persisted for decades in his thoughts.

Perhaps the simplest fact initially to be noted is that on April 6, 1917, when his country declared war on Germany, William Beebe was in his fortieth year. Hence he was not a likely candidate for the draft, nor was he a member of any peacetime armed force which might be called to active duty. Therefore any war service would be voluntary and in a branch of service of his own choosing. Theodore Roosevelt, Beebe's senior by nineteen years and a friend of long standing, volunteered to raise and lead a division on the western front. He could claim quasi-military experience in assembling his Rough Riders and victorious leadership in his famous (or vainglorious) charge up San Juan Hill in 1898. Despite these exploits, or perhaps as a consequence of them, Roosevelt was rebuffed by the military authorities and reduced to surrogate participation through his four sons. Beebe chose to volunteer at a far humbler level, and he was not refused.

But why choose war service at all? What business has a successful naturalist, now approaching middle age, at the battle front in France? Beebe's fellow ornithologist Frank M. Chapman was content to serve in the Red Cross; the older department leaders and staff members of the Zoological Society were willing, indeed eager, to muster a sort of home guard to defend New York City against attack. Not so William Beebe; for him it would be war service or nothing. His decision was the more radical one, as were the consequences also. But again: what impelled him to it?

Family tradition is a possible answer—one that might have inspired the men of Mary Blair Rice's family, for example, with a Confederate general to look back to. But little in the Beebe ancestry suggests warlike propensities; with the exception of the first John, who fought Indians in June 1676, and the first Roderick, who served at White Plains, military prowess does not distinguish the line. To be sure, William made an attempt to redden the escutcheon when he claimed that the third John had "taken an active part in two lusty Indian wars"—but Uncle Clarence, the family genealogist, sees this John as considerably less heroic: "He was possibly a private . . . from September to October, 1759, in the French and Indian war." So William Beebe could scarcely have found much warrant among his forebears for his wish to join the carnage on the western front.*

General war fever is another possible source of persuasion. Later generations must read in perplexity and dismay the accounts of public hysteria that come to us from those days. The congressional outburst on the occasion of declaring war was only the most official manifestation of the general loss of wits. Nor did Beebe need to go so far afield. The New York Zoological Society Park Guards, uniformed and armed, were presented with their company colors in May, 1917, to the accompaniment of these enthusiastic words from Henry Fairfield Osborn: "Our beautiful flag has a great history; no other nation has a flag which means so much; every color has a meaning. . . . We are behind the President of the United States, we

* An even more unlikely account of Beebe bellicosity (from sources unknown) appeared in the *New York Times* obituary for Charles Beebe, March 23, 1931: "Mr. Beebe was a veteran of long service in the National Guard. . . . He took part in the suppression of the riots here at the time of the Civil War." The riots occurred in 1863; Charles and Clarence Beebe were born March 13, 1852.

are behind the fighting mayor of this great city, we are behind our soldiers and sailors, and ready to join the ranks of the defenders of our city if we are so fortunate as to be called." Seven members of the Zoological Society staff were listed as serving in the regular armed forces that year, and thirteen in 1918.

If, as appears likely, Osborn was the second most influential person to touch Beebe's life, Theodore Roosevelt was doubtless the first. By the time war came to America, Beebe had known and worked with Roosevelt for at least ten years, from the time the President burst in upon the "nature faker" controversy with an interview and then an essay in *Everyman's Magazine* for June and September 1907. Beebe recognized in Roosevelt an expert field naturalist, a doughty champion of conservation, and an exceedingly persuasive person to enlist in support of his scientific endeavors. Apparently he also greatly admired Roosevelt the man; and one of the facets of this complex personality was infatuation with war. When his sons trained as officers and entered the service of the Allies long before most of their countrymen, Theodore Roosevelt hailed them for grasping the opportunity for "crowded hours of glorious life"—apparently his true notion of the realities of 1917, though his fine phrase came unacknowledged from Scott's *Old Mortality* of 1816. Quentin, his unmarried youngest son, was advised to ask his fiancee to come overseas and marry him, and not to worry about death or maiming: "As to your getting killed, or ordinarily crippled, afterwards, why she would a thousand times rather have married you than not have married you under those conditions; and as for the extraordinary kinds of crippling, they are rare, and anyway we have to take certain chances in life."[2]

Quentin had no great period of time in which to act on this paternal injunction. Presently he shot down a German plane near Château-Thierry, achieving, as his father exulted, "his crowded hour, and his day of honor and triumph." Roosevelt wrote this on July 12; on July 14 Quentin was himself shot down and killed. Nor was he the only casualty among the four: Theodore Jr. was gassed and took a bullet in the leg, Archibald was even more severely wounded, and Kermit was decorated for gallantry. "Yet if all our four sons should be killed," Roosevelt wrote to a friend on October 25, "their mother and I would feel that, even altho we were crushed by the blow, we would rather have had it that way than not have had them go."[3] The day the fighting stopped Roosevelt was seized by inflammatory rheumatism and hospitalized, returning home on Christmas Day. He died twelve days later.

Yet for all the force of Osborn's jingo rhetoric and Roosevelt's deathly infatuation, Beebe was not the mere creature of other and more powerful men. The impulse to enter war service was deep within, nourished by such confirmation but not created by it. Lord Byron had died in a brave little war for Greek liberation from Turkish tyranny—the perfect Romantic beau geste. But death, however excellent, was not mandatory. The literary heroes of Beebe's boyhood had no call actually to die; they fought with fine adolescent zest and daring in a variety of righteous wars, but, though wounded now and again ("I see your right arm is in a sling. What is your wound?" "Only a chop with a cutlass, sir.") they generally avoided expiring. These were the heroic striplings of George Alfred Henty (1832–1902), an editor of the British magazine *Union Jack* and a proficient writer of war-adventure books for boys. He wrote about eighty of these romances—and he was perhaps William Beebe's favorite boyhood author. In his fiftieth year Beebe could still recall "the long-ago Christmases when a newly-published Henty book was an invariable and almost the best gift." For years then the boy Beebe read of intrepid youths who dauntlessly attacked or bravely repulsed a variety of cunning enemies; who might suffer perilous reverses and yet win through; whose deeds were gallantly selfless, patriotically pure—and whose daring rescues might well involve girls of gentle blood, tender years, and delectable chastity.

To dismiss as trash such swiftly carpentered romances is both to question their expert appeal to boyish spirits and to belittle their pervasive consequences. And in the case of another of Beebe's favorite authors, Rudyard Kipling, no such denigration is possible. The Kipling fictions Beebe often refers to, *Kim* and *The Jungle Book,* and *Barrack Room Ballads,* a favorite book of poetry, are uncommonly expert productions, the work of a master of his craft. Nor is Kipling a bluff simpleton in the Henty manner; his ballads do not lack harshness and pain and doubt, and his Kim is a roguish young picaro whose ambitions are by no means totally fulfilled by his contact with war. Yet one may surely question whether the complexities of Rudyard Kipling were readily understood by a young man whose boyhood had been nourished so fully on G. A. Henty. At the end of one edition of *The Jungle Book* is a story more properly in keeping with those early years—"Servants of the Queen," wherein the soldiers of the Empire justly perform their sanguinary good deeds for its perpetuation. Poor Kipling, perhaps, but excellent Henty.

Finally, however, a man at forty is no more the product of books than of public opinion or influential friends. He is his own special self, the man that the years, early and late, have made him. "Under the provocation of extreme danger to me or mine, I have always valued human life at less than nothing,"[4] wrote Beebe in 1926, and in 1930 he stated that eleven men, whether part of his entourage or "native savages," had lost their lives on his expeditions.[5] All but two of these died during the pheasant expedition, when nine are said to have perished.[6] Here are claims and avowals that reach far into a certain region of William Beebe's true nature, and suggest more about one aspect of his spirit than his reading lists and acquaintanceships ever can.

But while such things speak of lethal impulses, they do not tell us, for example, why Beebe did not embrace the "splendid little war" of his own generation, eagerly rescuing the Cubans from Spanish tyranny or delivering the Filipinos from the selfish machinations of Emilio Aguinaldo in the best Roosevelt and Henty tradition. (The Cuban part, of course, was brief, and the Philippine aspect remote, protracted, and deficient in glamor.) Since then Beebe presumably had learned the thrill of firing from concealment at a native bowman and seeing that "miserable creature" roll down the hill dead or dying, his round hat spinning hoop-like beside him; and there were also those "native savages" whose deaths he speaks of but does not recount. In such killing a major psychic potentiality was released; within the complexity of Beebe's spirit apparently had grown desires which now could be satisfied only by participation in the grander lethality of a great war.

As noted earlier, the precise nature of Beebe's military service is not at all clear even to an attentive reader of his books, or even of his articles or the reports he submitted to the Zoological Society. One discovers that he flew over the battle zone in some kind of military aircraft, evidently a bomber—but when Roosevelt tells Quentin that "some flying and bombing" was permitted Beebe, but "not fighting the German war-planes," he imparts more exact information in one line than Beebe is likely to give in a page of war allusions. Furthermore, Beebe devotes far more space to experiences on the ground than to those aloft. He underwent aerial bombardment while in London, Paris, and Verdun. He saw his first German barrage from an automobile, and mentions Verdun, Furcy, and Dunkirk as sites of similar experiences; other scenes of battle and devastation are Fleury, Hadrecourt, Douamont, and Souville. He tells of combat patrols, Very lights, trenches, dugouts, gas. Yet commonly in all of this there are no names of officers or enlisted men, there are no dates, there is virtually no mention of units, nor indeed of Beebe's own direct participation, or even of his rank or assignment or duties. Even when he mentions briefly his actual flying, presumably on bombing or observation missions, and the sight and sound of antiaircraft fire, and the use of his parachute in a particular emergency, we do not know exactly where or when or in whose service. Nor is it helpful to turn to the lists of men called to the colors from the Zoological Society—Beebe is not included; similarly, brief biographical accounts yield few specifics, and Beebe's own entries in *Who's Who in America* offer nothing to the purpose. And so, regrettably and regretfully, one begins to suspect bogus claims and false heroics. Why otherwise the mystification?

The likely answer does Beebe neither honor nor dishonor. He told what he chose to tell, and presumably did not lie; he ignored what he wished to ignore, as not useful to his ends. Those ends were not journalistic or historical or scientific, but personal, literary, even in a sense fictional. He created his war story and his war persona in vignettes and asides, by artful selection—but again, there is no evident call to charge him with outright falsification. He wishes the reader to know that he was trained for aerial war, that he went to war, that he saw many things and reacted in particular ways. To establish these things he had no need to be explicit; implications may serve better than bald facts, especially if the latter are in short supply.

And if much is left out, it is for good reasons also. It serves no useful purpose for Beebe to reveal that he enrolled in 1917 not with an American or British force, but with the French Aviation Service, and spent the better part of his time in learning to fly and instructing other volunteers in the United States—nor that in the course of instruction he suffered a fall and incurred a severe wrist injury, which took him out of action for months. During his recovery period, however, Beebe made a trip to British Guiana to acquire a tapir, a spotted cavy, and a silky anteater for the Zoological Park. This was in September and October, 1917. Still recuperating, he then went to England and France on what an associate called "an observation trip."[7] Apparently, whatever the condition of his wrist, he was permitted to do a certain amount of flying and bombing, but his military status was anomalous—hence, perhaps, the prohibition against firing on German planes. Then Beebe left Britain by ship in time to visit Theodore Roosevelt at Oyster Bay in mid-March, 1918. Eight months of war remained, but William Beebe's overseas military career, if such it can be called, was over.[8]

The consequences, on the other hand, were only just beginning to emerge. In the *Zoological Society Bulletin* for May 1918 appeared three articles by Beebe, one directly on the subject of the Great War, the other two containing references to it. The main article was "Animal Life at the Front," telling of Beebe's observations "on various sectors during the past winter." The account is fairly straightforward, but even this early it is marked by imprecision—"various sectors," for example—and by unfortunate hints of self-congratulation or even dramatization. There is more than a mere hint of conventional propaganda in the contrast Beebe finds between the two sides separated by front-line trenches: one represents "barbarism," the other "civilization." Against this he poses the impartially sardonic definition of battle weapons and their work as "supreme efforts of mankind," and in the second article he speaks of the "temporary hell which man, with the aid of gas or high explosive, has been able to achieve." And finally, aside from the actual observations of animal life (with a few exceptions, unremarkable) there is a theme which pervades the article generally: revulsion and even horror. Levelly and at times mordantly, Beebe tells of "the *bruit* of the terrible struggle," of "wasted deserts and tortured landscapes," of "terrible conditions of sanitation and the numbers of unburied dead," of "raw ruins of farmhouses and villages," and of "the ghastly desert back of Douamont—a land of slime-filled shell holes, with half fallen wooden crosses and the flapping remains of old camouflage as the only relief from mud."

Still in 1918 Beebe wrote what became his best known statement on the war, the opening paragraph to *Jungle Peace:*

> After creeping through slime-filled holes beneath the shrieking of swift metal, after splashing one's plane through companionable clouds three miles above the little jagged, hero-filled ditches, and dodging other sudden-born clouds of nauseous fumes and blasting heart of steel; after these, one craves thoughts of comfortable hens, sweet apple orchards, or ineffable themes of opera. And when nerves have cried for a time 'enough' and an unsteady hand threatens to turn a joy stick into a sign post to Charon, the mind seeks amelioration—some

> symbol of worthy content and peace—and for my part, I turn with all desire to the jungles of the tropics.[9]

Initially the same notes are struck: front-line participation (with an undertone of heroics), war propaganda, dismay and revulsion. The new note is the frank recognition of personal jeopardy, fright, and escape. Swift metal may still shriek over the shellholes, the ditches may still be filled with heroes, but Beebe has grounded his aircraft and taken his leave. His wounds are psychic, not physical, but like Hemingway's Nick Adams, he has made a separate peace. If two paragraphs later he can still speak of "the high adventure of righteous war," he himself has nevertheless abandoned the heights and foresworn the righteousness of a war no longer his.

The theme of horror continues in Beebe's writing for many years. In *Edge of the Jungle* (1921) he moves from the sound of bees to "the blackness of night over the sticky mud of Souville [and] that terrible sickly-sweet odor of human flesh in an old shell-hole," and from the study of ants to "white, pasty mud which stuck to our boots by the pound . . . bitter cold mist which seemed but a thinner skim of mud . . . a sweet, horrible, penetrating odor"—all of this evoked by memories of a night near Verdun.[10] And so in many another book do references to war appear, as though forced unbidden into the light by malign pressures momentarily dominant. Here is a late example:

> In war men do all in their power to maim and kill one another, yet when wounded enemies are captured, instead of being subjected to the logical process of extermination they are taken to hospitals, cured if possible, exchanged for other cured prisoners and shot at all over again. I suppose it is the remains of the almost extinct idea of chivalry. Probably there are aviators who would hesitate at driving a sword through a woman or a baby who, without demur, will drop bombs on them when disguised as "civilian population."[11]

Something of a commonplace today, yes; it was beginning to be so even when Beebe published this, in 1942. Yet almost a quarter of a century had passed since William Beebe first began to speak out on the bitter realities of the Great War—and in America in those days his was a lonely voice and not part of a chorus. For example, the separate peace declared by Nick Adams provided one of the famous phrases of the literary reaction against the war, but not until Hemingway's *In Our Time* appeared in 1924. *A Farewell to Arms* with its scarifying scenes of war, its Caporetto aftermath—another separate peace by way of Lieutenant Henry's desertion—its memorable passage comparing war deaths to those of animals in the Chicago stockyards, even its reverberant title, was published in 1929; and the novel *The Sun Also Rises,* which derives so much of its power from the war wound of Jake Barnes (one of Roosevelt's "extraordinary kinds of crippling") appeared in 1926. Other Americans, to be sure, had spoken out somewhat earlier than Hemingway—John Dos Passos, e. e. cummings, Laurence Stallings and Maxwell Anderson, William Faulkner. Already a fairly well-known writer, Beebe chanced to be among the first to speak of war realities to a wide audience, and to announce the theme which other and greater writers would soon take up and elaborate.

None of this is intended as proof that William Beebe had a significant part, or was even a minor prophet in the development of a literature in reaction against the Great War. Although *Jungle Peace* began with a confession of desertion from a war still being fought by the writer's former comrades-in-arms, the book then concerned itself with Caribbean travel and jungle scenes and studies. The two passages cited from *Edge of the Jungle* had first appeared in *The Atlantic Monthly* a year or so after the Armistice, when the prevailing literary climate was symbolized by the award of the Pulitzer Prize to Willa Cather for *One of Ours,* an undistinguished and conventional war novel from a most distinguished writer; but again, *Edge of the Jungle* was about denizens of the tropics, not the western front. The serious and lasting literature of the war was yet to appear, and in this Beebe had no place, except perhaps as one who understood and spoke up before the others.

For all this, there is no reason to doubt the significance of the war to Beebe himself, or to anyone who wishes to understand him. It is important to keep his response at the level of meaning it deserves—personal, immediate, life-affirming. He was a man driven by unhappy compulsion to seek war, and to envision it through the eyes of others. But his own eyes, those of a scientist now and not a dreamer, revealed truths he accepted straightway, and his spirit made the erotic choice, for life, however common, and against death, however exalted. He had seen the sign post to Charon and turned the other way.

:4 Magna Opera: The Pheasant Books

> The work on the pheasant monograph is now practically complete with the exception of proof reading and book-making, and the final touches, which would have been completed immediately if it were not for the war.

By the end of 1914, when Beebe made this report,[1] a full five years of effort had passed, and still the great four-volume *Monograph of the Pheasants,* the chief scientific work of Beebe's life, was not finished. Seventeen months had been spent in search of the birds themselves; then much time had been devoted to study among museum collections of pheasant skins—for example, at Tring in Hertfordshire and in London, Paris, and Berlin, where Beebe spent three months in the summer of 1912. Arrangements for publication of the monograph had been made with H. F. Witherby and Company in London. The numerous color plates had been commissioned and painted; several were reproduced in Berlin by Albert Frisch, and others were in the process of being printed in Austria when the war began. By then the photographs to be used had been selected and integrated with the text, and the text itself, as Beebe indicates, was finished. But the fact that the color work was being done on the German-speaking side of the battle lines held up the project for the next four years.

The choice of a London publisher suggests certain parallels between Beebe's experience and that of John James Audubon nearly a century earlier. Audubon's great folios, even today the most magnificent ornithological works ever produced by an American, were too ambitious an undertaking for any American publisher to attempt. Hence the artist turned to the Havells of London to engrave and print and color his plates for *The Birds of America*, and to a publisher in Edinburgh to issue his text, *Ornithological Biography*. Again, when Daniel Giraud Elliot in 1872 brought out his elegant *Monograph of the Phasianidae*[2]—the first American work devoted to pheasants—he abandoned his earlier practice of doing his own art work, and engaged the Prussian-born London artist Joseph Wolf for his illustrations. He served as his own publisher but supported his project through subscriptions, and the great majority of his 114 subscribers were British—almost exactly duplicating Audubon's initial experience. (Among the relatively few Americans listed was "Theo. Roosevelt, Esq.," father of the future president.) That William Beebe turned to Great Britain for his publisher and to the continent for his art work indicates that such a lavish production was still more readily handled on the other side of the Atlantic.

In many ways Audubon and Beebe and their works differ rather widely. Beebe had little skill in drawing or painting, but an extraordinary talent for words, so he concentrated on the text and left the illustrations to others—G. E. Lodge, Archibald Thornburn, Charles R. Knight, H. Grönvold, H. Jones, and Louis Agassiz Fuertes. Audubon's genius with pencil and brush was by no means matched by his way with English (a second language) and so his text was edited for style by the Scottish ornithologist William MacGillivray. And in sheer physical size and artistic splendor, Audubon's plates far outstrip those of Beebe. The elephant folio page on which Audubon's paintings were reproduced measured nearly twenty-seven by forty inches, with many of his pictures virtually filling the page. Beebe's plates, regularly measuring eight and one-half by twelve inches, appear on a page approximately twelve by sixteen. And there is an inherent difference between a work ranging over the entire known avifauna of America, and one limited to the family Phasianidae; Audubon dealt with about 500 species, Beebe with only 112 forms, subspecies included.

Therefore the fair and proper comparison is not with Audubon but with Daniel Giraud Elliot. His two-volume monograph concerned the seventy-odd species of gallinaceous birds which he chose to include in the pheasant family, among them New World turkeys and African guineafowl. The text, seldom occupying more than a single page for each bird, was forthright and basically technical. Despite his use of enthusiastic objectives to describe various species, Elliot wrote to inform, not inspire. Hence his text depends for its lasting qualities on its accuracy and more especially on its currency. But it is too much to ask of a century-old work in natural science that it retain its place as a technical source; it is in the nature of science not merely to build upon the past but finally to bury it under mounds of new information, with only the work of the great discoverers still boldly visible. Daniel Giraud Elliot was no such discoverer. He did useful work in describing new forms, thereby making a place for himself in the ongoing ornithological enterprise; otherwise the text of his pheasant monograph has little value beyond the historical. The work's major interest today lies in its illustrations, the art work of Joseph Wolf.

Although size is not in itself a crucial factor, it is important in dealing with such large and often showy birds as the pheasants. Wolf had a full folio page to work on, and frequently extended his pictures to the very edge, even requiring a fold-out in one or two cases. He took full advantage of these ample dimensions, and knew better than to distract from his artistic effects with too much background. Perhaps he had learned from the splendid work of the English bird painter John Gould that vegetation and scenery can easily work against the major concern of bird painting, the bird itself; and he also paralleled Gould in presenting male and female birds in complementary positions well in the foreground, with all other features subordinated. It was with good reason that Elliot dedicated the monograph "to My Friend JOSEPH WOLF, ESQ., Whose Unrivalled Talent Has Graced This Work"—and whose enduring excellence, though scarcely unrivaled, gives the work its principal claim to later attention.

The illustrations for Beebe's monograph are that and little more. All are painted in detail, with as much of the appropriate setting for individual species as the artist could manage. (R. Bruce Horsfall provided various environmental scenes for background, but did not paint the birds.) Although their talents varied somewhat, none of the illustrators was the equal of Gould for artistic effects nor of Wolf for imposing individual figures. Louis Agassiz Fuertes, who demonstrated admirable talents in other works of natural history, does not appear to his best advantage in Beebe's monograph. Of the half-dozen illustrations he contributed, none reflected the subtlety and precision and sense of life that informed so many of his paintings of birds and mammals elsewhere, and especially his masterly work for *Artist and Naturalist in Ethiopia* (1936). So *A Monograph of the Pheasants* was not likely to make a permanent place for itself on the strength of its art work. But in interesting contrast, some of Beebe's photographs were quite distinguished, evoking scenes and habitats in a way the painters could seldom equal.

Beebe's great work therefore would depend for its major success on its worth as a contribution to science and its excellence as a piece of writing. The first, having to do with Beebe's innovations in classification, his discovery of a new species, and more broadly his field observations and life histories, will be taken up in Chapter 10. As to the writing, again a parallel with Audubon suggests itself. When he came to write *Ornithological Biography*, Audubon found he was unwilling to confine himself to birds alone, but instead ranged widely in a set of interpolations called "Delineations of American Scenery and Manners"—and even the bird accounts tended at times to abandon description for celebration. So also with Beebe, who found little merit in a mere array of facts. It was altogether apt that Beebe's friend and colleague Lee S. Crandall should begin the Beebe obituary for *The Auk* with a passage adapted from *A Monograph of the Pheasants*—a passage not of technical exposition but of descriptive prose richly evocative in its effect and almost cadenced in its language.[3] Nor was this excerpt unusual; scarcely a species is passed by without some vivid comment on Beebe's part, and many are favored (or burdened, depending on the reader's point of view) with pages of enthusiastic description or colorful narrative.

As a rule, for each species the text of the *Monograph* was divided into sections: "The Bird in Its Haunts," "General Distribution," "General Account," "Detailed Description," "Early History and Synonymy," and perhaps "Captivity" and "Relations to Man," plus an occasional section of a more specialized nature, for example on natural hybridization. Although a page or so may suffice for little-known species or races, Beebe gives nearly forty pages, for example, to the red jungle fowl, ancestor to our barnyard chickens, fifty to the other jungle fowl species, and more than twenty to the Bornean argus pheasant, about which he made valuable observations. Here and there, when his own data are lacking, Beebe quotes at length from others; but the deeply personal, intensely engaged quality of his pheasant adventures overrides such borrowings as easily as it obscures the technical data which underlie the entire work. Here in the major scientific publication of his career, Beebe uses the first person singular constantly, exuberantly, and with scarcely an apology. The pheasant journey to the Orient was a profound experience which would have to be imparted to the reader with all the eloquence at Beebe's command, even though the potential audience was quite small—the *Monograph* was first published in an edition limited to 600 copies and priced at $250.

It is generally under the heading of "The Bird in Its Haunts" that Beebe relates the circumstances of his own encounter with a given species, although his tales may appear elsewhere in the text also. The scientific accuracy of these passages may be assumed; the creative impulse is beyond question. Throughout the four big volumes, each running no less than 150,000 words, Beebe scatters his bird adventures in abundance, alert to the natural setting, the topography and climate, the special attributes of the bird he is pursuing, and the special problems attending the pursuit. In a rare moment of humility, he says at one point, "To make my few successful experiences fully appreciated I should write, in as vivid language as possible, accounts of my far more numerous disappointments and failures"—but in so saying, Beebe recognizes that his work does in fact celebrate the successes and recount the joys of his many months in the East. Not since Audubon had an American naturalist combined so zestfully the quest for birds with the stories arising from it; and never before had such mastery of language informed the telling. Something of this may have been in Theodore Roosevelt's mind when he finished reading the first volume of the *Monograph of the Pheasants* and, not many hours before his death, wrote to Beebe: "I have read through your really wonderful volume. . . . I cannot speak too highly of the work."[4] The relatively few readers of the *Monograph,* or of the abridged version called *Pheasants: Their Lives and Homes,* brought out in two volumes in 1926, were likely to agree with Roosevelt's laudatory estimate. No doubt they too found these works "wonderful"—but such readers were not great enough in number to satisfy a man who had become a writer both widely known and well rewarded, and whose creative demands were not yet fully satisfied.

The grandeur and sweep of the pheasant enterprise called for a popular book of like grandeur to encompass it. As a scientific expedition, the journey to the Orient produced the work from which Beebe's contribution as an ornithologist would largely be judged. Other consequences, marital, emotional, intellectual, also proceeded from this journey and awaited expression. Though the writing of the monograph was itself an imposing creative effort, much remained to be set forth. For more than a year the Beebes had moved as strangers in strange lands, seeking and finding things both sought and unsought. After more than thirty years, Mary was to write of these things only briefly and obliquely; but William was impelled to write at length of his experiences, confronting them as best he could, finding words for what had come to pass, telling what he had learned. Not at once, however; in contrast to all his other expedition books, *Pheasant Jungles,* published in 1927, was made up of chapters assembled over many years. The first appeared as a magazine article after four years had passed, the last only when the book itself was about to come out—and by then all the chapters, according to the author, had been rewritten.

If this procedure is unique in Beebe's career as a writer, so is *Pheasant Jungles.* It is his only major work of fiction—fiction not as dream or vision or fantasy, nor as the brainchild of novelistic imagination, but fiction as the creative reordering of experience, the manipulation of memory to do the bidding of art. It is doubtful that Beebe's complex spirit was ever free of such creative urges; he saw, correctly enough, that facts of the living world are multiform, hidden, even mysterious, and never fully to be apprehended. There is no more constant a theme in all his work than the vastness of the unknown in nature and the slightness of the inroads made by human knowledge. Nearly always Beebe is content to plant his small crop of data and to harvest from them what he can of understanding. In *Pheasant Jungles* his effort is of quite another sort. To be sure, he still works from data: he tells of a real land called Garhwal, real islands called Borneo and Ceylon, a known mountain called Kinchinjunga, and presumably real people named Hadzia, Mutt, Tandook, Aladdin. But his harvest is prefigured by the needs of his psyche and his art, urges and demands to which mere facts must bow.

At about the same time as *Pheasant Jungles* was being conceived and assembled, another American, much younger than Beebe and destined for far greater literary fame, was also reordering crucial experiences to produce a notable fiction. Through the first week of July 1918, Ernest Hemingway had ridden his bicycle on Red Cross duties on the Italian war front, had received on July 8 a terrible set of wounds from an Austrian mortar shell, had recuperated in a hospital in Milan, and had fallen in love with a nurse. Discharged from the hospital after the war ended, Hemingway went home, presently to be jilted in a letter from the nurse and later to write a book called *A Farewell to Arms.* In it a young American—now in the Italian army, not the Red Cross—begins a tentative affair with a nurse, receives the same terrible wounds on the same war front, resumes his affair and impregnates the nurse at a Milan hospital, is ordered back to duty in time to participate in the Caporetto disaster, and deserts to return to the nurse. They flee to Switzerland, she bears his child, dead, and herself dies.

Thus did Ernest Hemingway confront and resolve his unhappy experiences in love and war—by exaggerating almost beyond recognition his own part in the conflict, and by possessing the nurse and then, fictionally, killing her. William Beebe eliminated Mary Beebe by even more radical means in *Pheasant Jungles:* he simply caused her to vanish. The woman who accompanied this American naturalist on all his major ventures in the Far East through the year 1910 and on to May 1911 became, in the writing of the book, a non-person, a cipher. Her name is never given; her presence is never really made manifest; her words are never heard, her response is never known. She does not lie with her husband in any bed, she does not react to his desires, nor he to hers. Their dwelling may be a tent in the Himalayas, a government bungalow in Burma, a houseboat on the Pahang or a war canoe on the Mujong—or for that matter, a cabin on the *Lusitania* or the *Isis* or the *Lady McCallum;* but she is invisible, inaudible, intangible—a ghost.

But like any proper ghost, she remains to haunt the premises. As William regretfully watches the *Lady McCallum* out of sight, as he crawls into his sleeping bag on Sandukphu, as he listens to a chorus of gibbons in Malaya and finds it "as thrilling, as full of age-hidden memory meaning as the morning chant of the red howlers in the South American jungles," we sense her presence, barely repressed into invisibility; and when the expedition celebrates a snowfall in Sikkim, or when the Beebes are invited to a formal dinner at Kuala Lipis or to a tribal dance in Sarawak, the word "we" appears fitfully, almost materializing Mary before the vertical pronoun again prevails. (The painter R. Bruce Horsfall, likely to have been present at the snowfall scene, at that moment is also nearly rendered visible, for the first and last time in his six months of participation in the expedition.) On occasions more doubtful or forbidding, such as William's wretched seasickness aboard the *Isis,* his recurrent attacks of fever, and especially his times of nervous collapse, Mary's banishment seems rancorous, hostile, even malicious—a further traducing of the unremarked witness. We do not know how she fared at such times, nor what she did; we are told only that William, by his own avowal, faced them alone.

If there is denial and rejection in this, there is also the turning of basic sexual impulses and assertions into ostensibly nonsexual channels. So Beebe regularly lets it be known that he played doctor to the natives, assuming a role of authority and power sanctioned neither by proper training nor by cultural empathy and experience. So also he played judge and jury, summoning from time to time a miscreant member of his entourage before his judicial seat—a fair achievement for a thirty-three year old American naturalist, acting both as Sahib or Tuan and as a surrogate custodian of white rule. In fiction or reality, these are good declarative roles, and all the more so when William Beebe plays them as a lone white male far in an alien wilderness.

Traditionally, even braver assertions are found in deeds of valor; but daring deeds and participation in perilous undertakings are not for women. To scale the peaks, to ride the rim of mighty gorges, to penetrate the leech-ridden and leopard-haunted jungles, to travel by war canoe on savage rivers among the head-hunting Dyaks or by boat and afoot through the cholera-infested Malayan wilderness—these are masculine enterprises, and so to be related. Though Mary played an active part in all of them, the fact is simply not admissible, and serves only to becloud critical understanding. Upon the one great fiction of her absence the whole book turns; hence it cannot be judged except for what it is, a fictional work, a product of the creative imagination.

We have the author's own word on his fictional intentions as he tells us at the outset that he will employ "the oblique, corona-like visual and aural contacts" which years later may be "all the more vivid for being at the time semi-subconsciously received." That is to say, these are not so much recorded data as the stuff of memory, the peripheral impressions that persist after years have passed. And Beebe starts off his second chapter with an even more direct avowal, first by quoting his favorite novel of eastern life, Kipling's *Kim,* and then asserting that Kim and his saintly Red Lama are truly "incarnations from books to life . . . so much more real than most half-alive humans whose origin is other than in an understanding brain." Not merely should life follow art, as Oscar Wilde asserted, but art is truer than life, and *Kim* is Beebe's witness.

This novel, here quoted directly and many times alluded to, is especially significant as a touchstone for a man who was setting out to write his own fiction from experiences in the Orient. As more than one critic has suggested, the persuasive qualities of *Kim* spring in large part from Kipling's identification with his intrepid young scamp of a hero, fictionally reliving his creator's own childhood, but now with adventures unhindered and boyish dreams fulfilled. Nor is the nature of those adventures irrelevant to Beebe's effort; Kim, for all his native look and upbringing, is in fact a white boy, and in his rapscallion way serves his imperial Queen and helps to confound the knavish tricks of the imperial Russians, intent on infiltration and subversion of her realm. He plays his youthful part in the Great Game of counterespionage, vital to the maintenance of rule—but he is saved from Hentian fatuity by a sardonic boyish intelligence second only to that of Huckleberry Finn, and by his radical kinship (again like Huck's) with the lowly, the ruled and not the rulers. Kim is the lesser figure, perhaps because he can have it both ways—as Huck, lighting out for the Territory, knows he cannot. William Beebe was not the man to play Huck, but he could play Kim with relish and assurance.

The question of rulers and ruled is in no way avoidable. Except for the trip through Dutch Java, the Beebes spent all but their last few months within the sphere of the British Empire, experiencing for almost the first time (their relatively brief Guianese expedition being the actual beginning) the pervasive presence of imperial control. Mary in her brief account shows no clear reaction, but William in *Pheasant Jungles* offers sentiments of unmistakable approval. Among many words of praise for the empire and its functionaries, these are especially notable:

> I saw these splendid British men and women in the true perspective of their terrible isolation: their pluck in preparing for the oncoming fight against cholera, their holding true to the best traditions of their race. I remembered the justice and fairness of government of tiny Indian villages, of Burmese hinterland hamlets, of Sarawak Dyaks, and I thanked God that, next to my own country, I could claim blood-relationship with and loyalty to the other great English-speaking people.[5]

And these, concerning the enforcers of imperial control:

> As I came to know them, I developed greater interest and admiration for my Gurkhas. They were well-trained men, natural fighters, and the only mercenaries in the British army allowed to retain their native side-arms, the kokries. . . .[6]

> In the great Punjab and Northwest Provinces the Sikhs form a marvelous body of men. . . . The Sikhs form the backbone of the English native army and constabulary in India. When, as master, you win the respect and affection of a Sikh servant, you need fear neither poison nor steel in so much as it is humanly possible for him to protect you. . . . As one looks deep into their clear eyes one longs for a hint of their true ancestry. It seems altogether reasonable that their forefathers were the remnants of Alexander's Grecian army, many of whom settled in the northern provinces. And the kinship of face, of morals, makes of them companions beyond all other native tribes.[7]

Nor is Beebe himself denied a Kim-like role to play, however unwittingly, in the Great Game:

> When a stranger comes to a troubled country and goes about in jungle and over mountains unattended, intent only on stalking wild birds, it would be natural for any enemies to think that this must be some kind of trap, and that he had adequate means of defense not visible to the naked eye. I learned later that throughout the entire region we were known to be the advance scouts of the column of the 96th Punjabis who, with elephants, were later to push forward their punitive column to our present position.[8]

Perhaps, then, if "the entire region" around this camp on the frontier of China saw Beebe as a spy for a hated imperial power, a certain "renegade Chinese" saw him thus also—and learned that his means of defense were indeed adequate against a simple crossbow. In any case, to be taken for a spy and not a mere ornithologist is a good story, and one which would suffer ignominious diminution from the presence of another white scout named Mary in those same jungles and mountains.

If such pledges of allegiance to imperial rule had in fact marked the boundaries of Beebe's understanding, *Pheasant Jungles* would be a book so bereft of insight as to be fit for shelving alongside the memoirs of middle-level colonial administrators and lieutenant colonels who served anywhere east of Suez. But if Beebe seemed never fully to confront the morality, the politics, or the economics of empire, he never failed to respond to its human aspects, in particular the social and the ethnic. William Beebe's loyalty to the imperial idea—whether here in British dominions or years later in American-occupied Haiti—was conventional and in no way surprising, given his upbringing and the circles in which he moved. But his was a spirit neither conventional nor truly elitist, nor was it lacking in complexity, faced by the vastly complex East. Special edges and angles of perception saved his book from intellectual banality, much as deep emotional engagement bestowed upon it artistic depth and range.

In books of science one looks askance upon generalizations not backed by necessary facts convincingly arrayed. In fiction or poetry it is rather otherwise: "Happy families are all alike"; "It was the best of times, it was the worst of times"; "Heaven lies about us in our infancy!" Beebe himself made fun of the sentence which declares, "The French are a gay and polite people, fond of dancing and light wines"—but then he was not writing fiction. Now in *Pheasant Jungles* he tells us that Tamils are "a progressive, sturdy, diligent people, traders by instinct," and that "the Cinghalese do not adapt themselves quickly to new conditions [and] are too gentle by nature to offer any serious resistance to any advance" (although elsewhere he states that "the Cinghalese give no ground and do not stand aside"). Similarly, the reader is assured that "strong emotion has no lasting place in a Hillman's mind," while the Chinese in upper Burma "are the most stolid and at the same time the most emotional [people] in the world," and "wherever there is Mongolian control there is also mystery and secrecy. Hidden motives lie behind the smallest trivialities of the day. Men pass by on the trail like shadows, and their faces tell nothing of what is in their hearts." On the other hand, the mountain Tibetans "are immune to suffering and privation; their excess of jubilance and joy in living spills over in the midst of the hardest labor. They laugh at everything, good or bad"—even if on the next page we find mountain Tibetans who are "stolid, unwashed [and] stupid, with that impregnable stupidity which far transcends the reputed stupidity of animals."

Not merely for their patent contradictions, but even more for their unblushing certitude, such bald statements must trouble the conscientious social scientist, or the careful observer of any sort. But such a person has no business here; his testimony is irrelevant to the point at issue, the success or failure of Beebe's fiction. Generalizations on these exotic peoples are well taken if they provide the color and excitement and strangeness desired, ill taken if they do not; and for the most part they serve their author well. As the journey northward from Darjeeling into the matchless Himalayas derives some of its joy from the high spirits of the Tibetan porters, so the expedition into troubled Burma gains a sense of ambiguity and threat from the "mystery and secrecy" of "Mongolian control." So also the splendor of the Hill and Hills is enhanced by the presence of the daring but pathetic Hillman, himself so unlike Beebe the western scientist, and yet, like him, caught in the spell of these glorious steeps. Similarly the Tamils and Cinghalese and Dyaks of Beebe's stories lend vivid dimensions to the work—and this without regard to the actual nature of the people themselves, or their true relation to William Beebe. To create people out of memory and imagination and need is to perform the fiction writer's necessary task; and to persuade is to prevail.

With character sketches in his earlier books Beebe went far to convince us that he was no common colonialist, interested merely in "faithful gun-bearers" and other indigenes properly trained and subservient. He gave us such clearly individualized natives as the trial defendant Ram Narine, the child-bridegroom Madhoo, and the Akawai Indian hunter Nupee, whose non-Christian marriage Beebe considered honorable enough, and whose status as a subject of "protection" by the white conquerers seemed unhappily anomalous. More than once, furthermore, Beebe expressed regret and dismay at the aping of white masters by colonial peoples. Now in *Pheasant Jungles* we find several character studies which even more forthrightly call into doubt aspects of the imperial presence. In Ceylon Beebe encounters and sympathetically portrays the fate of an engineer educated in the West but doomed as a mere native to perform tasks far below his training and abilities. In Garhwal he affectionately describes the young Hillman Hadzia, and finds himself momentarily resenting the intrusion of a company of British soldiers upon this wilderness scene, where "Tibetans and Hillmen . . . for centuries had preceded them and for many years would follow." In Kuala Lumpur he finds much to reflect upon and admire in the story of a young Chinese whose intelligence and persistence and thrift enable him to learn the photography business, set up his own tiny shop, and at length buy out his original Malay employer—a symbol of the Chinese penetration which goes forward peacefully and inexorably in the East, whoever presumes to wield the power of rule. And in Borneo among members of a certain tribe Beebe sees the aimless copying of European articles of apparel as both ludicrous and sadly emblematic of the colonial power to which these unfortunate natives are capitulating—"links in the first chain forged by a distant civilization."

Yet it is pointless to seek in Beebe the contemplative irony of Joseph Conrad or the satirical perception of young Herman Melville or the wit and seeming detachment of Somerset Maugham. In any given locality Beebe's stay was fairly brief and usually dominated by professional concerns. At nearly every point he depended on official favor to provide ways and means and people and even escort for his expeditions. He was not unmindful of these things, nor ungrateful; and if at times he found himself challenging on intellectual or moral grounds the imperial venture, he scarcely made of these insights a consistent stance. The profound ambiguities which emerge when one culture confronts and dominates another; the destruction which attends any form of conquest, military, economic, even religious; in short the unlovely realities that underlie the lofty imperial avowals—these were not William Beebe's study, nor, except in scattered instances, his basis of understanding and response. Instead wherever he went his response was to "the people, and what they were making or were endeavoring to make out of their lives." From such deep interest emerged those separate persons who in the looking back became characters, sharply delineated and memorable, in the fiction he had set himself to write.

And as with character, so with incident. Since he did not employ straight chronology or geographical sequence, Beebe had no major organizing principle for his work, divided as it necessarily was among diverse scenes and nations great distances apart. He depended on separate excursions and adventures in each area to give his book the qualities he strove for, analogous to those of Kipling's novel. *Kim,* in common with other picaresque tales, has a wide geographical range and many separate exploits, and is loosely held together by two themes, the Great Game and the quest of the Red Lama for his sacred river. Beebe's quest for pheasants, his single major thread, took him even more widely over the eastern world, and each locality provided him with stories to tell. And although he had no Great Game to play, except unwittingly, he had a great desire to impart to readers his thrilling and perhaps perilous moments, his arduous efforts, his rewards and delights—and with all this, perhaps to reassure himself that his tour in the Orient was a spendidly successful undertaking and not the unhappy journey that in some ways it doubtless was.

Beebe's adventures in Ceylon began mildly enough, with his landing by outrigger canoe on the beach and his journey by ox cart into the interior. Yet before many days had passed he had survived two threats to his life. Only a violent shove by his Tamil tracker saved him from the lethal fangs of a Russell's viper, and only his swift climbing of a tree took him out of range of three huge water buffaloes which suddenly rose from the high grass and charged him. He tells of these things briefly and even with touches of humor, eschewing the kind of heroics that both vitiate the force of such events and call for even more breathless accounts in later chapters. Here in his first chapter Beebe suggests a shrewd awareness of the need to shape his book in a pattern of rising action within which these encounters, however perilous, must serve more as early enticements to the reader than as major adventures; and in fact he honors this pattern by taking his two most exciting and violence-ridden chapters out of chronological sequence and placing them for dramatic effect in the middle and at the end of his book.

Leaving Ceylon for Calcutta, Beebe offers portents of grander things as he takes the train northward from that city through strange and varied scenes to Darjeeling, the gateway to the great eastern Himalayas. He foregoes any attempt to portray the majesty of their incomparable vistas, but cheerfully describes the pandemonium which attends the departure of his expedition into these same mountains. Quickly then the reader is transported to a densely forested Himalayan glade, vividly depicted and abounding with life. Hints of drama are given by the appearance of deer and bear and pheasant and owl, but the only exciting event actually described is a sudden hailstorm, murderous in its consequences for many kinds of forest creatures. The one major adventure that comes to Beebe in these mountains again holds the threat of death, or at least seems to: he slips on icy crust to the very edge of a sheer drop of 1,500 feet, saving his own life at the last moment by finding a foothold which stops his slide; and then he discovers that the edge of the precipice gives onto a broad ledge just a few feet below, with a mantel of soft snow upon which he would have landed quite safely. All this is cleverly done, with the reader by turns apprehensive, relieved, and amused—but at no time allowed to forget that just below that fortunately located ledge there is indeed more than a quarter of a mile of empty Himalayan air.

Both setting and incident contribute powerfully to the drama of the next chapter, laid in the magnificent Hills on the border of Kashmir. It is night, but Halley's comet sheds its strange glow upon the great forested slopes where Beebe and his Hillman crouch and listen. Again there is danger abroad, abruptly signalized in the terrified bleat close at hand of a stray sheep as it is struck down by one of the great cats. Then another drama of life and death begins in the darkness with the pursuit of a young flying squirrel by a pine marten, both creatures dropping almost on Beebe's head from a tree above, and a third participant, the mother squirrel, arriving through the air to scream in helpless rage at the deadly pursuer. Only the flashing of Beebe's light causes the marten to give up the chase; otherwise the young squirrel would have perished. And in the midst of all this, two koklass pheasants have been frightened off their roosts and have gone squawking away in the dark. Then the Hillman is sent back to camp to sleep, and Beebe, in a nicely composed diminuendo, watches through the rest of the night, seeing the comet fade and vanish and the new day arrive. Later comes a placid interlude, a kind of ornithological detective story, but the chapter ends with a hailstorm even more violent than the earlier one, in which Beebe himself is endangered and his horses are driven frantically to cover.

Although it is far out of place chronologically, "Wild Burma" was chosen by its author to occupy the center of his book, physically and perhaps creatively. It is the longest of his chapters and the most crowded with agitation and danger. After suffering through a period of acute nervous affliction, Beebe again heads northward into the mountains, but now with troops as his escort, and with knowledge of a much larger imperial force soon to invade and punish certain rebellious border tribes. Beebe tells of killing one renegade and firing at others who threaten him; another native dies, possibly as the result of Beebe's surgical meddling; a third is killed less than a mile from the expedition camp, "probably by leopards," in the words of Beebe's photo caption; and a fourth, an insane Kachin boy, disappears after several days into the forest, doubtless to perish from starvation or leopard attack. Tigers are also to be found nearby; shortly before Beebe arrived at the village of Sin-Ma-How, the headman had lost three mules to them. And despite the fact that snakes are rare in this region, Beebe had the deeply frightening experience of seizing the body of a king cobra as he slipped and fell blindly in a hillside thicket. Fortunately, according to his story, the snake responded only by crawling away.

Thus briefly outlined, "Wild Burma" sounds like Henty at his most lurid, or even like one of the penny dreadfuls that Beebe seized upon as therapeutic reading in the midst of his breakdown. How well he weaves these melodramatic elements together, how shrewdly and knowingly he underplays the drama and understates the intrepid qualities which may so readily be inferred, how expertly, in short, he carries it off, is a measure of his skill—as a writer of fiction. Nowhere would Mary's presence have been more disastrous, nowhere is her banishment more crucial. And nowhere is the basic cleverness of Beebe's peculiar stance as an author more admirably demonstrated. May he be charged with inventing implausible characters and involving them in unlikely actions? On the surface, surely not; he is a scientist, indeed a noted one, who merely observes and records, though blessed with perceptions and writing skills not commonly found among professional birdmen. But may he then be assailed for bogus science, even nature faking? Scarcely; if tigers and leopards and king cobras are found in northern Burma, who can say with certainty just where these creatures were and what they did, within the range of William Beebe's awareness in the late months of the year 1910?

Although in the next chapter among Beebe's "Servants and Super-Servants" there is the Sikh syce Naraing Singh, evidently alert to save his master from steel or poison (and whose first ghostly appearance at the campfire causes Beebe to reach for his gun) and the Chinese cook whose excellent meals are in no way compromised by the fact that he has been convicted and sentenced to a life term for poisoning half a dozen people, this chapter fits into the loose pattern of the book as a humorous interlude. The regular cook, called Mutt, turns out "epicurean sauces" but goes mad at each full moon. Umar the Malay boatman, instructed to return downriver to Fort Kapit in Sarawak for additional money, instead produces from "the small folds of his waist cloth" as many as 200 Straits dollars in silver coin. No indication of the weight of such a waistline treasure trove is given, so perhaps the virtue of the story lies in whimsical exaggeration. And when Naraing Singh is told of our Civil War, and asks whether General Pickett charged with elephants at Gettysburg, Beebe gravely suggests that the thickness of the jungle prevented it.* Accompanying such light moments are reflective comments on the special nature of the master-servant relationship, particularly in the story of Aladdin, the young Cinghalese Malay whom Beebe employed as his personal servant for a good part of the eastern journey.

A brief encounter with a highly venomous sea snake, brightly patterned in blue and red, begins Beebe's sixth chapter; and before many pages he is deep inside a vast Malayan cavern with only a flashlight to guide him through this lightless labyrinth. At length, in one of the deepest recesses, Beebe encounters "tragedy—fitly staged in this black hell." A bat by some mischance has broken a wing, and now as he lies writhing is being attacked by "two horrible gnomes . . . a long, sinuous serpent, white from its generations of life within the cave, and a huge centipede, pale, translucent green, sinister as death itself. I shuddered as I beheld this ghastly tableau,—serpent and centipede both emblematic of poisonous death, preparing to feast upon a yet living bat, devil-winged and devil-faced." The Irish fabulist Lord Dunsany, the author whose name appears most often in Beebe's writing, and to whose work he constantly alludes, is here accorded the ultimate praise of imitation, and successful aping it must be called, although the story breaks off without a proper ending, almost as if Beebe had grown weary of his own fancy.

Soon Beebe is far off in the jungle in pursuit of pheasants never seen alive in the wild by white men, and tells of his excursions with a directness and economy worthy of Edgar Allen Poe. "Late one afternoon," one account begins, "I reached a steep land-slip which, a few months before, had carried away a wide swath of jungle, leaving the disintegrated rock exposed or decorated with the new-sprouted plumes of yellow green bamboo. I had had a long, tiresome tramp, and was two miles from camp, across a deep, dark valley." He removes "the usual unpleasant collection of leeches" and, concealed behind bamboos, sits down to watch. Drongos and hornbills fly about and alight and fly again, while Beebe still quietly watches. Then two tree shrews engage in a fierce tussle in the branches

* This story, perhaps based on the exchange of letters between the King of Siam and President Lincoln, is one of Beebe's less original humorous inventions. Any battle-keen Sikh, informed of Pickett's charge over terrain too densely grown for the employment of elephants, would surely then ask, "In jungle of such thickness, O Sahib, what *did* this Pickett Sahib charge with?"

overhead, soon falling close by nearly exhausted. With these events as preliminaries, the climax comes swiftly: three bronze-tailed pheasants, and then two more, appear at the edge of the scar, their attention attracted by the fall of the shrews. The birds remain in clear view for a brief period, and then, to sweeten the victory, one adult bird steps into a shaft of light from the setting sun and fully displays its gorgeous plumage to Beebe's enchanted gaze. As the sun fades the birds disappear, and though Beebe hastens in pursuit, he sees nothing more of them.

Gibbons call far away in a rollicking chorus; dusk is coming, and with it a storm. Turning quickly toward camp, Beebe crosses the shadowed valley as the night creatures begin to stir about him. "The darkness settled down as I reached my hammock, emphasizing the many spicy jungle odors and ushering a wind which rattled the bamboos and shook every loosened leaf to the ground." In this spare fashion the story ends, and the three separate searches for the ocellated argus—each, for its own particular reason, a failure—are told in a manner similarly disciplined.

This chapter, called "From Sea to Mountaintop in Malaysia," first appeared in the *Atlantic* early in 1918. The succeeding chapter, "Malay Days," came out nine years later in the *Zoological Society Bulletin*. The book itself was about to appear, so "Malay Days" may have been written as a collection of further adventures designed to round out the Malayan story with new material. In any case it is a very different piece of writing, far more laden with human drama and tension, and with aspects of calculation and contrivance not so plain in the prior chapter. "Malay Days" begins and ends with tales on a fairly hackneyed theme, the nerves of white men broken by tropical tensions. In the first instance the afflicted ones are British colonials and their wives at the torrid outstation of Kuala Lipis; in the second the victim is Beebe himself. While the former, telling of an oppressive dinner party, is more of an incident than a full story, the latter is an account of several pages. Both are expertly manipulated, and on first reading convincing. Doubts begin to stir with another look at the events at the dinner: a disputed statement which caused a full five minutes of tense silence; what ensued when Beebe finally broke the silence with a casual remark —"all leaped hysterically to answer"; the indecision of a woman who resolved "to clear out for Kuala Lumpur" but who changed her mind "three times in two courses." Then in the second account one becomes aware that Beebe is using a fairly implausible story of three lepers (presumably hired in the black of night) as a literary device, along with other melodramatic situations to persuade the reader of his nearly demented state. And near the close of the story Beebe offers a startling anachronism to augment an emotional effect: still fighting his profound agitation, he shoots a peacock which falls "in a veritable tailspin which awoke shuddering memories." In the first sentence of *Pheasant Jungles* he had mentioned the same tailspin, and added "two wing-slips"—but unless we may assume that William Beebe learned to pilot an airplane between the time of Kitty Hawk, late in 1903, and December 1909, we must assign his experience with slips and tailspins to his wartime training, many years after he had left the deck of the *Lady McCallum* or the jungles of Malaya.[9]

The last chapter of the book is called "With the Dyaks of Borneo." It is relatively long, achieving some of its length through passages adapted from Beebe's early papers on the flying lemur and the pangolin, and a description of the veritable menagerie of creatures Beebe set up in his jungle camp. Its main purpose is to show why, as Beebe had reported sixteen years earlier, the trip with the Dyaks "in a seventy-foot canoe far up into the interior [was] in many respects one of the wildest and most interesting of our explorations." Having since changed the possessive from "our" to "my," Beebe gives us scenes of wilderness savagery and danger to his heart's content, with recurrent emphasis on the primitive and warlike nature of the Dyaks themselves. All of this reaches a climax of sorts in the story of Drojak, one of Beebe's paddlers, who is attacked while far from camp and alone in the jungle by twelve men of a hostile tribe, but who emerges untouched from the battle, with four of his opponents dispersed in flight and the other eight killed and decapitated—the weapon used being the very sword hanging in Beebe's studio as he writes. All eight heads Drojak ties carefully to his belt, "taking great pains with each knot," and thus ponderously adorned he walks back to camp. Forthwith he is dubbed Drojak-no-spear-can-touch-him, and his reenactment of this Homeric encounter was a rare treat at subsequent campfires.

Drojak himself, tall and muscular and appropriately unmarked, is pictured in *Pheasant Jungles,* but the manner of disposal of his trophies is not given, and Beebe spares us photographs of them, individually or collectively. But other such trophies, black and desiccated and bearing eyes of white wood, hang in a circle over Beebe's head in the Dyak "great house" where a tribal dance featuring his paddler Narok takes place, described by the author in the final pages of his book. These heads seem to watch over the wild, almost hypnotic savagery of the dance itself, imparted to the reader in vivid and compelling language. *Pheasant Jungles* ends with Beebe's departure from the scene downriver in his great canoe; it is late at night, and as the author looks back he sees the lights of the house suddenly go out, leaving only blackness between.

To analyze the manner in which this book was conceived and put together and to assess the implications of the result is a study both fascinating and troubling. It was noted earlier that Beebe garnered special advantages from playing a dual role of putative scientific observer and fictional romancer. In one way, however, the consequences of this role remain thoroughly ambiguous. The issue is not simple fakery—such outright nonsense as Kipling employs in *The Jungle Book,* for example, when he has the huge python Kaa batter down a heavy masonry wall with what is in fact a relatively fragile snout and head. Something more subtle and more pervasive is involved: the use (and misuse) of dangerous creatures not in faked activity or fierce encounter but as threats, casually stated or merely implied, to Beebe's life.

"With the Dyaks of Borneo" begins with Beebe swimming in the Mujong River after a long jungle tramp. "I clung to a half-submerged vine and let the current sway me back and forth, and searching with my eyes for a chirping insect on an old fallen tree near by I suddenly saw close to my face a six-foot serpent coiled on a branch which still kept its bark above water. I did not recognize it, but it was manifestly a 'hot snake' as my Dyak interpreter called poisonous species. . . . I was swimming amid the shadows of a strange tropic river, with a venomous serpent watching me. . . ." In a fashion not unusual with him, Beebe goes on to deprecate the danger which he has just set forth; but let us note the expert and yet dubious manner in which this naturalist has persuaded the reader that such danger in fact exists. His finding the snake is convincingly circumstantial and fortuitous—he was looking for something else. The snake, of the impressive length of six feet, he does not recognize; but it is "manifestly a 'hot snake' " and hence one of the "poisonous species," and two sentences later it is flatly "a venomous serpent." This is not so much fakery as verbal trickery, whereby an unidentified snake becomes a threat to Beebe's life without any of the proof a reputable naturalist must give—and at the same time, without the caution that a good writer of fiction, concerned not with scientific accuracy but with plausibility, would show toward so glib a passage.

"Disregarding the rumors of tigers and black leopards, I crept through the jungle in the dead of night"—so begins another of Beebe's tales, in this case relating a hunt for the argus pheasant in Malaya. There exist not merely rumors, but *the* rumors, and not just one apiece of these great cats, but plural numbers. Beebe rather favors the plural in this connection: In Burma the headman's mules die in an attack by "tigers"—an encounter the man later describes in pantomime—outcast youngsters succumb to "starvation or leopards," a strange tribesman is probably killed by "leopards" and is found "half devoured by wild beasts."[10] The reader of *Pheasant Jungles* discovers that in every chapter except "Servants and Super-Servants," one or both of these deadly species can be found—deadly by clear implication if not by the more common fictional device of an outright struggle, man against beast.

Finally there is the matter of the poisoned arrows allegedly shot at Beebe's camp near Sin-Ma-How in Burma. The poison used, Beebe says, "is aconite and sometimes tetanus germs." Aconite, described as "fatal" in the second chapter, is a plant (in this case probably *Aconitum uncinatum*) bearing a substance making it dangerous or even lethal when eaten by mammals—but is that substance also deadly when injected on the point of an arrow? As to tetanus, its pathogen, the bacillus *Clostridium tetani*, is deadly indeed, and injection by arrow would appear to be as effective a method as any "renegade" could wish; but if it is "sometimes" used, what about this particular instance? Beebe had a considerable collection of spent arrows (he states that all the arrows fell harmlessly from his tent canvas, with one exception which stuck; none had the impetus to pass through) and plenty of time later to discover in the laboratory whether the bacillus was in fact present. At issue, whether in fiction or reality, is the death of a man, a man who may have been shooting arrows tipped with lethal tetanus, or with a substance derived from aconite, or possibly with neither; we are not explicitly told.

One may wish to know from a sense of moral disquiet, but even more from the feeling that Beebe's immunity as a romancer must give way here to his responsibility as a scientist involved in a deathly encounter. The data he provides are insufficient and their implications equivocal, and in matters of such gravity, nothing is settled by terms like "miserable creature" or "renegade" for one's adversary. At a very different level, it is also a dubious proceeding for a naturalist to declare an unknown serpent venomous merely on sight, or to affirm, subtly or with obvious intent, the man-killing and man-eating potentialities of "leopards"—do they hunt in packs? —without satisfactory proof, firsthand or other. Nor are the implications of ubiquitous peril fearlessly risked, bestowing upon Beebe a kind of jungle machismo built of knowing reiteration, especially happy ones.

What compounds the problem—and the disquiet—is the testimony given by the original printed form of Beebe's tales. The first version of most of his expedition experiences appeared in *A Monograph of the Pheasants,* the next of which was finished by 1914. Nearly all of Beebe's ornithological adventures and certain other events in *Pheasant Jungles* are adapted from passages in the *Monograph,* some essentially word for word, others in altered form. Several of the changes are intended, harmlessly enough, to exaggerate success: the "tailspin" peafowl is shot and taken to camp, whereas in the monograph version all ten birds escaped and the day's hunt was fruitless; in *Pheasant Jungles* the display of the Malay bronze-tailed peacock pheasant is splendid and full, rather than so brief and partial that Beebe "could not appreciate the beauties thus displayed"; the finding of Sclater's impeyan is a triumph because "no white man had ever seen this bird alive before," although both Elliot and Beebe in their monographs mentioned a bird of this species brought alive from Assam to a London zoo, where it was still living in 1872. Other changes are intended to stress the peril of the author's surroundings: what earlier had been "the eternal, mournful, four-toned call of a hawk cuckoo" becomes the falsetto bleat of a sheep "struck down by a leopard or a tiger"; the sight of a supposed water buffalo—actually a rhinocerous—had caused Beebe to sit still and watch, not jump for a tree and swing himself up, as he later wrote in *Pheasant Jungles;* and as to the "hot snake," it was simply absent from the original passage in the monograph.

Other things of greater importance were absent too, whether from the monograph or from other prior versions. In May 1916 Beebe published in *Harper's Magazine* an article called "Pagan Personalities," from which he derived passages to appear later in *Pheasant Jungles,* especially in the chapters "Servants and Super-Servants" and "With the Dyaks of Borneo." There are odd contradictions between illustrations in the article and the book, and even greater divergencies in the two texts. Narok and the dance at the great house are described in *Harper's,* but Drojak does not appear, whether as Beebe's foremost native assistant or as the collector of human heads virtually *huit d'un coup.* A community house of hostile Dyaks is described in almost the exact words used eleven years later in *Pheasant Jungles*—but the book adds this passage: "My own Dyaks were not wholly at ease as we paddled by on the opposite side of the river. . . . [With my field glasses I] raked the whole building. In a score of places I could see armed Dyaks peering through the chinks at us, and near one end were the rifle-like muzzles of three blow-pipes pointing our way, with whose poisoned arrows I was already acquainted."[11] No elucidation of this final cryptic comment is offered here or elsewhere.

Even as late as 1927, when "Malay Days" was printed in the *Zoological Society Bulletin,* almost none of the dramatic details of the formal dinner can be found, nor is there any celebration of "these splendid British men and women" as given a few paragraphs later in *Pheasant Jungles.* And the other story of shattered nerves, Beebe's own, was totally lacking in the original monograph version, where the day is described from beginning to end as benign, moderately eventful, and unproductive: "My quest this particular day was futile, the birds were too much on the alert, and a wretched little babbler set up a screeching alarm just as I had settled into a good point of vantage in an ancient buffalo wallow, and the Peafowl did not stand upon the order of their going."[12] Which is simply to say that there were no lepers, no random gunshots, no attack of fever, no hysterical response to the leap of a squirrel or the approach of leeches, no fall of a peafowl in a tailspin nor of William Beebe into a terrible maze of thorns. The two versions are not merely different, they are contradictory; yet both describe the same day.

The strangest and most disturbing of such contradictions concerns the danger to Beebe's life posed by certain natives near Sin-Ma-How in Burma. "On two of my trips after pheasants I had rocks rolled down on me," Beebe tells us in *Pheasant Jungles,* and relates how on the second occasion a shot frightened off his assailants, "two miserable Kachins." The same paragraph begins the tale of the native bowman whom Beebe shot dead. There can be no doubt as to the place in question, an area described as a watershed in the *Monograph* and so designated (in precisely the same words) in *Pheasant Jungles.* The expedition had, the *Monograph* states, "an escort of six Gurkhas. The Kachin tribes hereabouts are nominally safe, but the individual components of these tribes are uncertain quantities . . . [but] I worried little about human enemies and only twice was even threatened with any molestation."[13] Presumably this statement accounts for the two occasions involving rocks—but it leaves the native bowman story in doubt to an unsettling degree. "Molestation" is far too mild a term for an attack (on four successive nights) by poisoned arrows; yet Beebe's concluding sentence is altogether explicit, and buttressed by the emphatic words "only," "even," and "any."

But then comes the version of events at the Sin-Ma-How camp printed in the *Harper's* article mentioned above: "For some time marauding bands of Chinese and mongrel tribes had been further complicating the situation along the frontier, so that the entire country was in a state of unrest and upheaval. As a matter of fact we were never seriously molested beyond some minor skirmishes with Mongolian robbers fleeing inland for safety. Once, at night, they shot down on our sentry with poisoned arrows, but these did no damage beyond striking the walls of the tent and knocking down whatever happened to be hanging up there." In this telling there are several assailants, not one "renegade," but the attack occurs on one night only, not four, and Beebe is not recorded as killing anyone.[14] So despite Beebe's *Pheasant Jungles* photographs of his supposed assailant at the Sin-Ma-How camp before his death, and of the man's crossbow afterward, did the attack with poisoned arrows occur elsewhere? Or did it occur at all? The same question can be asked of other events in *Pheasant Jungles* which are absent from earlier accounts or which contradict them, or are given with important alterations.

The web is a tangled one, and perhaps at a later time, when the full record is made available—as presently it is not—each strand can be traced and the complete pattern made known. The crucial judgment, however, will still be literary and not moral. Ernest Hemingway gave us several fictions derived from his Italian war experiences, of which *A Farewell to Arms* is the most complete yet scarcely the most accurate; but it is creatively the fullest and best and psychologically perhaps the most revealing. William Beebe gave us one version of his pheasant adventures in the *Monograph*, certain further versions in magazine articles, and his most fictionalized final version in *Pheasant Jungles*. In this book he also achieved his most interesting creative synthesis, and in so doing offered the most telling implicit comments on Beebe the man that he would ever commit to print. The dismay one may feel at seeming fakery or dishonesty is understandable enough, but no more valid as a critical standard than similar feelings directed against Hemingway. One may also draw unhappy inferences from the character and intent of Beebe's changes as he writes for a mass audience—inferences which are of real use in understanding the complex and, here at least, troubled spirit of William Beebe, but otherwise of little utility.

Near the end of the Drojak story, Beebe compares this battle "to the last fight of Umslopagaas," and at the climax of the tribal dance he reverts to Kim and the strange spell cast on him by the mysterious agent Lurgan. These are useful, indeed essential references. A mythic tale and a novel of boyish romance and imagination are explicitly cited when, here at the end of his most ambitious popular work, William Beebe offers two tales of wild Dyaks which he clearly intends as a powerful concluding flourish to his fiction. By these signs we are reminded both of the author's fictional stance and intention, and of his hope to match means to ends, fictional techniques to fictional purposes. At the start of *Pheasant Jungles* Beebe gave the reader fair warning, and repeated it at the conclusion. Perhaps he hoped thus to guide readers and critics in the proper direction, that they might judge the book which in fact he wrote, not the one which, for all the superficial objectivity of his settings and scenes and photographs, he did not write.

For if at last a book of fiction beguiles, entertains, informs, if it offers the reader a conviction of enlarged experience and heightened understanding, then it has done its proper work, and needs no other excuse. Such as a book in its special way is *Pheasant Jungles.* To it William Beebe gave his most engaged and extended creative labors and his most profound emotional commitment. He did not do so in vain; the book offers one passage after another of splendid power, and a cumulative effect of elevation and sweep he achieves nowhere else. The compelling qualities that mark all good fiction, the sense that scenes, events, relationships, responses are as they should be and not otherwise, Beebe affords in abundance. And if his work does not convince at every point, nor always cohere in time and place and theme—a fate common to tales of the wanderer—yet it shows throughout a powerful impulse, the need to assert the man in his risking death and surviving, in his taking of life and yet also affirming it. From this the book *Pheasant Jungles* derives its subtle resonance and its unique fictional success—a success which William Beebe could claim through major creative effort, and Mary Blair Beebe through her spectral presence, giving to that effort a special urgency and force.

:5 The Tropical Jungle

> The terrors of serpents, tropical insect scourges, and other [jungle] dangers of which we had been forewarned, existed, so far as our experience went, entirely in the minds of our friends in the North.
>
> *Two Bird-Lovers in Mexico*

The experience William Beebe mentions here was that of sharing a tropical camp with Mary in 1904 about twenty miles from Manzanillo in the state of Colima, and more than four degrees south of the Tropic of Cancer. It was the first time either of them had ever lived and wandered in the kind of hot lowland forest commonly called a jungle, and they both thoroughly enjoyed the adventure. Great spiders abounded, hundreds of bats swept low through their campground at dusk, all kinds of strange noises came out of the darkness, and once a ten-foot boa slid silently past the tent in the moonlight—all of them part of a natural scene the Beebes had come to explore, and, to such eager young seekers, gratifying rather than alarming. And here they found trogons, motmots, parrots, woodhewers, boat-billed herons, and many other tropical birds they had hoped to see, and would find nowhere else on their Mexican journey.

Next came their search for a wilderness in coastal Venezuela and Guiana, and then their last expedition together in the Orient, much of it in tropical regions considerably more trying than those of the New World. But William did not lose his enthusiasm for the tropics—nor, initially at least, did Mary, who continued to explore and write of tropical areas for several years after their divorce—and before long Beebe had shifted the emphasis on his career from birds to tropical research.

That the jungle is a place of beauty and wonder and general benignity is a theme found in all Beebe's jungle books, and an argument he offered against the proponents of the common notion of "green hell." Beebe himself had reflected this notion when he said in an early essay that "the fever-stricken wanderer in tropical jungles listens to the sweet notes of birds amid stagnant pools"—a rhetorical flourish which he was later to reprehend, but of the sort that other writers had employed for centuries, and would continue to use, if only to thrill the credulous and sell books. Thus one recent writer tells of shooting an anaconda more than sixty feet long (thereby doubling the record accepted by herpetologists) and reports another running twenty feet longer. He also suggests that this snake's noisome breath has the effect of drawing its intended victims into range, and then paralysing them.

Another modern authority, in page after page of hothouse prose, informs us that the Guianese jungle is full of unceasing trials and dangers, a place of slime, rot, loathesome and deadly creatures, disease, affliction, torment. His persuasiveness is much reduced, however, when he states that the ponderous capybara has an otter's agility in the water; that the sloth's blow, swift as lightning, can rip out a man's guts; that a kind of "python" (presumably the boa constrictor) grows to about thirty feet and the anaconda to nearly sixty—while among Guianese birds there is a stork bigger than an ostrich, a toucan so awkwardly formed that it can fly only short distances (the hoatzin, perhaps?), "trogans" of nearly fifty species and hummingbirds of five hundred.

As a popular writer as well as a scientist, Beebe did his best to refute such misbegotten effusions. He did not, however, deny the reality of painful or threatening tropical situations when he found them, for example in the equatorial lowlands of Brazil, or in the leech-ridden jungles along the Pahang in Malaya. Nor did he belittle the danger posed by certain tropical snakes; he had perilous encounters with Russell's viper, the king cobra, the fer-de-lance and the bushmaster, any one of which could have taken his life. Among the mammals he had particular respect for the Asiatic water buffalo. He was forced to escape these great beasts half a dozen times by climbing trees, and to shoot one of them when it threatened to knock him from his shaky perch.

It was also clear that William Beebe had no desire to emulate his friend Theodore Roosevelt by venturing into a jungle wilderness truly untrodden by white men. The famous River of Doubt exploration came about when Roosevelt, already well along in the planning of his trip through northern Paraguay and southern Brazil, was told of this unmapped region and immediately responded, "We will go down that unknown river!" So on February 27, 1914, he set off down the river with twenty-two men in seven dugout canoes and with food for fifty days. Roosevelt got malaria almost at once; boats were smashed in terrible rapids, one man was drowned and another killed by a crazed fellow boatman, equipment and food supplies were lost. Already wasted by fever, Roosevelt was further weakened by short rations, dysentery, abscesses, and a prior leg injury, so that finally he asked that the rest go on without him. Necessarily they refused, though by then Roosevelt needed virtually to be carried. In mid-April the first signs of white civilization appeared, and presently the wretched expedition came to the first town.[1]

The contrast with the journeys in *Our Search for a Wilderness* is sufficiently plain. For all their brave talk of "untrodden wilderness" and "jungles untamed," the Beebes passed through regions which had been known to whites for decades and even centuries, and which had already been exploited for natural products of various kinds; and they were never far from settlements. So it would be also when in 1916 William Beebe established his first Tropical Research Station; the jungle was indeed nearby, but so also was a major city, easily reached on a regular schedule by steamer.

In the spring of 1915, having resumed his regular tasks after a five-year leave in pursuit of pheasants, Beebe visited the zoological gardens of Pará (now Belém) in Brazil, to secure a collection of forty-three species of birds for the New York Zoological Park. While in Brazil he took time to investigate the jungle, concentrating on a single area and one huge tree, a *canella do matto.* The tree yielded seventy-six kinds of birds, while "four square feet of jungle débris" were found to contain more than five hundred organisms. Beebe reported his findings in *Zoologica,* the *Zoological Society Bulletin, The Atlantic Monthly,* and finally in *Jungle Peace.* Here was clear evidence of a change in his scientific approach; instead of ranging widely over the world in lengthy expeditions, Beebe had turned to a more static and concentrated form of study, one which he had mentioned approvingly as early as 1906 in *The Log of the Sun*, but which he had neglected in the intervening years. Now he was preparing to resume it, and seeking a proper site.

The Pará region was no such place. Lying directly on the equator, this swampy jungle was indeed something of a green hell, with temperatures frequently ranging above 100 degrees, daily rains and high humidity, mosquitoes and ticks and *bêtes rouges*—and "the mood of the jungle," one of "uniform, sombre mystery." For all the abounding life of this region, Beebe wrote, "it was death—or the danger of death—which seemed in waiting, always just concealed from view."[2] If a research station were to be established for jungle studies, it would have to be in surroundings considerably more benign than he discovered here in the Amazon lowlands.

Returning to New York at the end of May, Beebe was soon in contact with Theodore Roosevelt, who, despite the recent wilderness experiences which had permanently impaired his health, took up the idea with much of the old zest. At Oyster Bay the two men discussed various aspects of the proposed station, and Roosevelt doubtless exerted his considerable influence with the Zoological Society on behalf of the plan. What opposition he faced is not clear, but it was obvious that the erstwhile Curator of the Department of Birds, who had already spent much of his tenure in absentia, was about to assume a new role, with birds only one of his many concerns. Beebe's expanding interests were shown in the composition of the party that left New York in January, 1916, to choose a site and establish the first station: with Beebe as leader, it consisted of G. Inness Hartley, embryologist, Paul G. Howes, entomologist, Donald Carter, collector, and two artists, Miss Rachel Hartley and Miss Anna H. Taylor.

From Georgetown in Guiana the members of the party spread out in search of the best possible location, following several leads that came to nothing. Then in March they were offered a site which they were quick to accept— a great house called Kalacoon, formerly the residence of the Protector of the Indians, located on a wooded hill overlooking the Mazaruni River, about two and a half miles from the village of Bartica. The English planter overseeing the property simply told Beebe to occupy and use the place if he wished, and with this act of generosity the expedition came into possession of its new home. The main room to the front, raised high like a ship's bridge and facing the trade winds, became both laboratory and men's dormitory, thirty feet wide and twice as long, with sixteen windows.

As is often the case with older wooden buildings in the American tropics, there were plenty of apertures admitting not only air but many native creatures as well: scorpions, tarantulas (which kept down the roaches), and vampire bats, many of the latter actually roosting in the building and flying softly around the great room at night. In the absence of flies and mosquitoes, the party used no nets, but kept a wary eye on the vampires, and collected them or studied them alive at will—an excellent example of the benefits of a fixed and comfortable station for scientific work, with subjects for study not only close at hand, but actual residents of one's own dwelling.

In setting up the station for living and working, Beebe and his party initially faced a crowded schedule, since their first official visitors, Colonel and Mrs. Roosevelt, were soon to arrive. But Georgetown, in those days a city of 60,000, provided plenty of household furnishings and utensils, as well as ice, mail, and fresh food supplies every other weekday. Hence their distinguished visitors could be properly welcomed, comfortably housed, and well attended to. Roosevelt had indeed come a long way from the River of Doubt (soon named Rio Roosevelt in his honor) where he had struggled so painfully two years earlier.*

In such favorable surroundings Beebe had easily enough achieved his initial purpose, which was to banish the notion that all scientific work in the tropics must be both arduous and full of peril. But it still remained to be seen whether the efforts of Beebe and his staff at Kalacoon would produce worthwhile results. As soon as they returned to New York at the end of August they set about preparing for publication the data from their six months of study. In January 1917 the Zoological Society issued a substantial volume called *Tropical Wild Life in British Guiana,* written by all three scientists and well illustrated with their photographs and figures, which persuasively answered the question.

"There is no doubt," said a reviewer for the British ornithological journal *The Ibis,* "that far better results in the matter of

* Nevertheless Roosevelt felt that at least one major improvement to Kalacoon was in order. Writing to Osborn soon after his return, he asked whether the Zoological Society could find somebody to donate Beebe "a Ford car" to help him get about the countryside. Roosevelt then published a lengthy report on his visit, "A Naturalists' Tropical Laboratory," in *Scribner's Magazine,* January, 1917.

collection and observation can be obtained by working from a fixed centre, and that this is so is clearly proved by the present volume." Beebe's introductory chapters described the area around Kalacoon and something of its earlier history, and then the setting up of the Tropical Research Station itself. His main scientific contributions to the volume concerned birds—hoatzins, toucans, gray-backed trumpeters, and tinamous, plus "Ornithological Discoveries," a chapter written in collaboration with Hartley. In habitats ranging from cleared areas through scrub and second growth to the deep jungle, 281 species of birds had been found and many studied in detail.

Toucans, those oddly engaging birds of the American tropics whose great bright bills seem so absurdly disproportionate to the rest of the creature, had been known for centuries but never fully investigated. Around Kalacoon Beebe found five different species, and his chapter devoted to them is the first proper account ever written about nest sites, eggs, and young of these birds. Similarly, Beebe closely studied the ground-dwelling tinamous—tropical forest counterparts to quite remotely related partridges and quails—and discovered among other things the different roosting habits of the two genera found in the region. Partly he did this by drawing the correct conclusions from study of the tarsi of different specimens—and partly also by asking the opinion of his Akawai Indian hunter Nupee, who, as it turned out, had known the basic facts long before Beebe and the other Zoological Society palefaces appeared in his native jungle.

The featured ornithological chapter of *Tropical Wild Life* concerned a bird studied not at Kalacoon, but instead along the Berbice River near New Amsterdam, about fifty miles southeast of Georgetown. This was the hoatzin, a bird strangely arrested in its evolutionary development, and showing what Beebe elsewhere called "classic reptilian affinities." In company with his wife, he had first seen the hoatzin in March 1908 along the Guarapiche River in northeastern Venezuela, and again in April 1909 on the Abary River east of Georgetown, where Beebe took photographs of quite respectable quality, considering his equipment—a Graflex with a cumbersome 27-inch lens. At neither place, however, had he been able to study the young birds, and these provided him with his most remarkable data in 1916.

Strange though the adult hoatzin is, with plumage that seems permanently disheveled and awkward flight and frog-like voice—

and of course its history of survival in the present epoch, restricted as the bird is to an almost impossibly narrow habitat—the young bird is even stranger. Beebe chanced upon a colony where the young were all about a week old, covered with scanty black down and showing only half an inch of flight feathers. They were not precocial as are the chicks of many ground-nesting birds, running and feeding as soon as they are dry, but neither were they sprawled and nearly helpless, as are week-old squabs. Instead they were fully alert to the perils imposed by a party of inquisitive scientists, and altogether self-possessed in meeting them.

The first hint that he faced a worthy opponent came to Beebe as he approached a hoatzin nest—a ramshackle platform of sticks lodged about fifteen feet above the water—and managed to persuade the brooding bird to take wing and blunder off. Immediately there rose above the nest rim a nestling's head, in appearance as much reptilian as avian, with big intelligent eyes and a beak resembling that of a young tortoise. The head stretched up and up on a long, thin neck; presently, as Beebe's assistant climbed into the nest tree and drew nearer, the little bird stepped without haste over the rough twigs to the nest edge, using wings almost as arms for balance. With the assistant creeping closer, the hoatzin nestling climbed up as far as he could on an adjacent limb, using not only his feet but the remarkably formed thumb and forefinger on each wing for purchase.

Finally the bird could climb no higher, and the man still advanced. When six feet separated them, the little hoatzin dived headlong into the water straight below and disappeared. It took minutes of looking to discover that he had swum under water to emerge many feet away, cautiously sticking out only his head and neck along some floating debris before pulling himself from the water. Again using both his feet and his mitten-like wings, the little bird climbed back through vines and branches to the nest from which he had so bravely departed perhaps a quarter of an hour earlier. As Beebe says, an equal feat for a human child would have been to dive from a height of 200 feet—and also, he might have added, swim an ever greater distance underwater and then clamber back through a tangle of rough limbs to the starting point.

This was only the first of a fascinating series of observations and experiments with hoatzins, described at length in *Tropical Wild Life* and in briefer form in *Jungle Peace.* As with the other species studied in detail, Beebe was here making original contributions to ornithological science and thus demonstrating the validity of the tropical research station idea. Equally convincing on this point were the detailed biological studies made by G. Inness Hartley and Paul G. Howes, whose researches taken together made up the large portion of the book.

What is not so clear is Beebe's professional trend and purpose. As in the case of *The Bird* a decade earlier, *Tropical Wild Life* was a book which combined close observation and analysis with a goodly amount of fine writing, the latter notable for the improvements which the years had bestowed. Once again Beebe had written a book, or part of one at least, with a fundamentally professional bias, despite its excellent literary qualities—in brief, a book as much of science as of scientific adventure. Particularly as it was the first major publication to emerge from his new Department of Tropical Research, *Tropical Wild Life* posed the question of Beebe's own direction.

The answer came the next year with *Jungle Peace,* and would be reinforced by the three other books Beebe subsequently published on his experiences at research stations in tropical South America. Already, in fact, the increasingly literary trend of Beebe's career was well established. He no longer depended on newspapers and recreation magazines for access to a wider public; by the time *Jungle Peace* came out he had published many articles in *The Atlantic Monthly* and a few in *Harper's Magazine* and *The Century.* That is to say, he had reached as high as the nature essayist could go, and had joined the company of Thoreau and Burroughs and Clarence King and John Muir, all of whom had appeared in these august periodicals. In so doing he had met exacting standards and had been vouchsafed a distinguished readership, facts he was not likely to scorn; nor was he unaware that the market for good popular essays and books was a lucrative one. Like *The Bird,* therefore, *Tropical Wild Life* was essentially an aberration, and the questions it raised about Beebe's career were largely illusory. At least in ornithology, his principal area of scientific competence, Beebe had now made his major contributions and written his last book.* A mere glance at the chapter titles of *Jungle Peace* and its successors —ranging from "Sea Wrack" to "Parade of the Maggots"—is proof enough. For the rest of his life William Beebe would be a scientist

* As noted in Chapter 4, the volumes of *A Monograph of the Pheasants,* virtually completed by 1914, did not begin to appear until four years later.

still, but more than that, a watcher and wonderer and teller of nature's manifold tales.

In the foreword he wrote for *Jungle Peace,* Theodore Roosevelt shrewdly and rather bluntly set this book apart from Beebe's "earlier and more commonplace work." One may quarrel with his criteria at times, but scarcely with his general thesis. The charm of language which Roosevelt recognizes—and which he compares, aptly enough, to that of Robert Louis Stevenson—is not new, but it is more artfully employed. There is nothing novel in the fact that Beebe is a trained observer, but he has learned to wear his scientific learning lightly. Nor do we need to be told that Beebe has traveled widely and adventurously, in light of his earlier books. The true differences are of form and approach, and they do indeed set *Jungle Peace* apart from the Mexican and Venezuelan books and from *The Log of the Sun.*

The basic form Beebe had followed earlier was the chronological narrative, with digressions of limited scope only, complementary passages of description or analysis, and asides of reflective comment or personal response. The reward of this form generally lies in its clear sequential pattern, whether for writer or reader; the penalty lies in a certain monotony, partly induced by the undemanding nature of the pattern itself. The tendency is to let facts and events speak for themselves, with the result, in the case of *Two Bird-Lovers in Mexico* and *Our Search for a Wilderness,* that they may chatter to the point of tedium. Roosevelt's word "commonplace" is perhaps too strong; in the least of Beebe's books there is a distinctive quality which is finally the mark of an arresting personality. To give this personality free play, to learn the best devices for its expression, to escape the common narrative and descriptive conventions and the confining demands of chronology, had been William Beebe's successful study since his last book in 1910.

In *Jungle Peace,* however, his escape from chronology is more real than apparent. The opening paragraph quoted earlier suggests a return of the troubled bomber pilot to his tropical forest refuge, and succeeding chapters support the illusion by telling of travel southward among the islands of the Caribbean toward Guiana. Then the next to last chapter tells of a new area, equatorial Brazil, and ends with the author on board ship, presumably returning to New York after finding peace of spirit in his jungle wanderings. But in fact this is a chronology falsely imposed and supported by unobtrusive interpolations.[3] Beebe returned to the United States from the war front in March 1918, and, by his own report, was prevented by wartime restrictions from going to Kalacoon and working there. From internal evidence, the voyage being recorded occurred in 1916, with some parts perhaps taken from Beebe's brief trip to Guiana in September and October 1917. Furthermore, the return by sea he describes occurred in 1915, at the end of the Pará visit. So Beebe was still making concessions to the appearance of a proper time sequence, even though he had actually abandoned it.

His whole approach to writing had in fact markedly changed. In *The Log of the Sun* he had been a prisoner (though a fractious one) of the months and seasons; in the books he wrote with Mary, of the geography traversed and the data provided by each separate region. By and large he wrote then as a scientist, observing and recording in ordered patterns. But in many passages in his *Monograph of the Pheasants,* and in magazine articles which later became chapters in *Pheasant Jungles,* Beebe had proclaimed his freedom from such narrow bounds. Again in *Jungle Peace* he approached his materials creatively, not merely in discrete passages of response and interpretation, but in the broader plan of a given chapter, by shaping and ordering his materials for chosen ends.

Such creative freedom, however, can betray as well as bless. In the hands of common nature writers it may lead to mere fakery, a vulgar misdemeanor Beebe almost never committed; but in the hands of a natural scientist it can engender something a good deal worse, a stretching or heightening or twisting of observed facts beyond permissible bounds. Although the boundary line may often be dim and jagged, nevertheless it is there; and from time to time there would be those who suggested or flatly stated that Beebe had transgressed it.

Just as it is misleading to try to follow the bogus chronology imposed on the chapters of *Jungle Peace,* it is also unprofitable to think of these chapters as constituting a formed and plotted book. Except for the first chapter of two pages, each one had appeared separately somewhere else. Eight of the eleven chapters came out in *The Atlantic Monthly,* mostly in the first half of 1917. Aside from the fact that they were all written by the same man and concerned his own experiences, they could claim no more kinship than was afforded by the place of original publication, one of America's foremost general periodicals. Whatever unity this provided was of an editorial kind, based on the magazine's standards of literary performance. But if Beebe wrote well enough to capture the fancy of *Atlantic* readers, he was under no obligation whatever to stick to ornithology or any other biological science. Rather the reverse, in fact; straightforward natural history was not especially that magazine's style. Beebe was following a market, and doubtless his own preferences as well, when he wrote the pieces which eventually were gathered into a book: expertly wrought familiar essays, felicitous in style, broad in range of subject matter, humane in outlook, genially personal in tone.

As Roosevelt says, Robert Louis Stevenson comes to mind; and he might have added William Hazlitt and Charles Lamb and perhaps Washington Irving. None of these was a naturalist; all were basically men of letters, and experts in the art of the familiar essay. Stevenson, for example, brought out a volume in 1882 called *Familiar Studies of Men and Books*. Beebe's *Jungle Peace* might have been more accurately called *Familiar Studies of Men and Other Diverting Creatures*. Indeed, for the first few chapters a reader knowing nothing of Beebe himself might be mildly puzzled about his calling: is this man some sort of biologist with an avocation as a writer, or is it the other way around?

"At the earnest of winter—whether biting frost or flurry of snowflakes—a woodchuck mounts his little moraine of trampled earth, looks about upon the saddening world, disapproves, and descends to his long winter's sleep." This passage, Thoreauvian in its wry empathy and spare strength, argues the creative writer who knows something of nature. Later aboard ship the author collects sargassum weed for study and traps minute forms of life called plankton for microscopic analysis, thereby suggesting marine biology as his central concern. Soon, however, the ship reaches the island of Martinique, and the author tells of his dismay at "insufferable tourists" who spoil that lovely place for him—until it is redeemed by a splendid black woman, seen from shipboard as she emerges from the dead ruins of Saint Pierre: "Even at this distance I could discern her stately carriage, swinging and free, her black countenance and her heavy burden. At the very edge of the water she stopped, lifted down the basket piled with black volcanic debris and emptied it. She stood up, looked steadily out at the passing steamer and vanished among the shadows of the ruins." No more than a flourish in the manner of Conrad, perhaps—or is this man in fact a writer of fiction, traveling to gather materials for short stories or a novel?

Author Beebe appears to give the game away when he says early in the fourth chapter that he has come to set up a research station for a zoological society in New York. This avowal convinces less and less as the chapter proceeds through a series of Guianese vignettes to the trial of one Ram Narine, a coolie laborer who had assaulted another East Indian with a club and a rock. In the account of these austere judicial proceedings there is deft characterization and human interplay conveyed with a nice feeling for nuances, and there is implicit comment on the ambiguities of white rule over a colony of dark-skinned peoples. At the end of the chapter the reader may well wonder what happened to that research station mentioned twenty-four pages earlier. And so with other chapters; in the midst of "A Hunt for Hoatzins" Beebe takes us on a delicately evocative and gently nostalgic tour of the library, now long neglected, of the colonial club in Berbice, and in "A Wilderness Laboratory" he spends a full third of his pages describing in vivid detail a coolie wedding. That Beebe carries such things off with zest and skill is beyond dispute—but finally, to what purpose? To meet the forthright demands of his market; to demonstrate his virtuoso writing abilities; to prove his wide interests and intellectual attainments; to enlist public sympathy and support for his work; to argue the worth of his studies, the correctness of his approach, the validity of his conclusions; to impart to all who will listen the joy and wonder of his quest and of the world in which he pursues it. The reasons are complex, as was the man. In a dozen books he was still to write, all were operative in one or another proportion; and in the process William Beebe would become the best known American naturalist-writer of his time.

The year 1918 saw Beebe assume the title of Honorary Curator of the Department of Birds and take up full duties as Director of the Department of Tropical Research. Though wartime conditions did not permit his return to Kalacoon, the wartime market for rubber persuaded the owners of the surrounding estate to cut and burn the second growth and plant new trees. Thus the ecology adjacent to Kalacoon was considerably altered and much of the charm lost. When in 1919 the Zoological Society staff returned, it was to move the station to a new site called Kartabo, about three miles up the Mazaruni River at its junction with the Cuyuni. The large bungalow in which they set up the laboratory and library and mess facilities had once been part of a gold-mining operation, now long abandoned. It sat in the midst of a grove of tall bamboos, exotic trees brought centuries earlier from Java by the Dutch, and its wide front veranda looked across a small clearing to the river. Initially the bungalow served as quarters for Beebe and his two assistants, but later the staff lived in tents beyond the main buildings.[4]

Again the prospects for study were richly diverse—as Beebe put it, "birds and monkeys . . . great butterflies and strange frogs and flowers"—and the professional people broad in their interests and attainments. The fixed and relatively comfortable nature of such a research station also permitted Beebe to add to his staff several young women, beginning with artists to provide illustrations for articles and books, and culminating eventually in female colleagues with whom he shared research and publication. The artists at Kalacoon had been Miss Hartley and Miss Taylor; initially at Kartabo they were Miss Isabel Cooper—who had accompanied Beebe and others to Guiana in 1917—and Miss Helen Damrosch, who married Beebe's foremost assistant, John Tee-Van, in 1923. Once more the idea that roughing it in the midst of alien jungle scenes was a purely masculine activity was denied; these young women, well proven by their work to be talented and by their photographs prepossessing, put all such assertive male notions pleasantly to rout.

Whereas Kalacoon had engendered *Tropical Wild Life,* a work of more than 500 pages intended to be the first volume of a projected series, Kartabo, where the period of study extended over many years, produced technical papers for publication mainly in *Zoologica,* as well as articles for the *Zoological Society Bulletin* and more general periodicals. The fourth volume of *Zoologica* (March through July, 1925) carried "Studies of a Tropical Jungle; One Quarter of a Square Mile of Jungle at Kartabo, British Guiana." Here in almost 200 pages Beebe gave a description of the site of his studies, an account of the white man's presence in the area from the sixteenth to the twentieth centuries, notes on some of the flora, and an annotated list of the entire range of faunal life. Thirty more pages were devoted to "The Variegated Tinamou," giving a life history of the bird, a study of its hyoid and syrinx and the voice here produced, and various measurements which had been neglected or given inaccurately in studies by earlier scientists. Then, as with *Tropical Wild Life,* the larger portion of the volume was given over to papers by others—close entomological studies by Maud D.

Haviland and Alfred Edwards Emerson, and a detailed description of the isopods collected at Kartabo by Willard G. Van Name.

Perhaps the most interesting of Beebe's technical studies from the Kartabo years was "The Three-Toed Sloth," which came out in *Zoologica* in 1926. Here he applied his talents for field observation and experimental work not to a bird but to a strange tropical mammal. The resulting monograph is one of his most absorbing and persuasive scientific efforts, and it indicates again something which was becoming plainer year by year, the broadening of Beebe's professional interests beyond his early calling of ornithology.

As Kalacoon had provided much of the material for *Jungle Peace,* Kartabo produced two more popular books of essays, *Edge of the Jungle* (1921) and *Jungle Days* (1925).[5] Since most of the chapters appeared originally in the magazines noted (again, mainly *The Atlantic Monthly*) the two books can be considered together. Any attempt at a unifying chronology has been abandoned, and both the war theme and the scenes of Guianese life have been reduced. In this sense the two books are narrower in range; in the scope of natural life studied they are broader. Several of the chapters concern insects, ants in particular; aquatic life predominates in others; plants, from microscopic forms to great jungle trees, receive much attention; vertebrate creatures of diverse kinds appear, with birds and mammals only two classes among many.

So if *Edge of the Jungle* and *Jungle Days* can be said to have unifying themes, they are of the broadest—ecology and evolution. Increasingly the fixed site for study provided by Kartabo, the continuity offered by many months of residence over a period of several years, and perhaps most of all the infinite complexity of the tropical environment itself, impelled William Beebe toward these disciplines. He does not pretend nor perhaps even aspire to some resounding synthesis; he admits from first to last that his "scanty crop of facts" is planted in a "huge bed of ignorance," but he continues devotedly to cultivate his garden. Always there is more information to discover, more understanding to seek, and the fascination of the search never ends. In 1915 Beebe could be found absorbed in the study of a few square feet of jungle debris; ten years later, his zest undiminished, he climbs and crawls and stoops and peers through the vast tangle of a fallen jungle tree, seeking day after day the secrets of its nearly limitless life.

Over these years the reader of Beebe has been required to follow new paths and more intricate arguments. As the author has become more intellectually venturesome, more inclusive, perhaps more trenchant and challenging, he has also become geographically fixed, his adventures localized and his human contacts narrowed. Beebe's grace of language, his sensuous gifts and nice wit (and always that complex, vivid intelligence) unfailingly persist to beguile the reader —who may well need such comfort as he makes his way through the intricacies of ant life, for example, or the ecology of a jungle beach. In a sense Beebe has returned to the finding and recording of facts as in the Mexican and Venezuelan trips, but with even greater concentration and more encompassing scientific knowledge.

A comparison here between Beebe and Thoreau is tempting but should not be pressed beyond useful limits. Both, for example, tell of ant battles; but where Thoreau is ironic, detached, philosophical, Beebe is engaged to the point of empathy, even as he remains the empirical observer. However, Beebe's conclusion has a Thoreauvian edge to it: "such terrible unthinking altruism" as ants display in their endless repetition of communal functions finally cries aloud for individual assertion, even for "selfishness and crime"—anything to relieve the implacable round. One wishes for more such comments, in themselves providing another kind of relief, from the tyranny of facts. With so rich a fund of precise data, Beebe was able too seldom to escape far enough from his own learning to reflect upon it, as Thoreau did so cogently; and so at last, in many of the chapters of these two books, the reader learns too much to understand.

But too many exceptions appear to be readily ignored. In *Edge of the Jungle,* "Hammock Nights" shows Beebe at his witty best, combining a certain grave ostentation with the sharpest of responses—a playful and knowing performance. "The Bay of Butterflies," concerned with an inexplicable migration of hosts of yellow butterflies from the shelter of the jungle to the shore and on out to sea, finds Beebe both intensely engaged and convincingly mystified by so strange a death ritual. In "Sequels," the last chapter, Beebe reflects on events of past seasons and brings some of them up to date, discovering new aspects to old perplexities. The strangest eventuality is the fate of a colony of army ants, perhaps the same legions which had briefly taken up residence in a Kartabo outbuilding eighteen months earlier, before vanishing into the jungle. Now they have returned, and this time are put to rout by chemicals wielded by human beings. The surviving ants march off in a disciplined column into the undergrowth, eventually curving back and forming a great closed circle—around which they blindly scurry, night and day, until nearly every ant has perished from exhaustion. Here Beebe draws no moral, and doubtless needs none.

Jungle Days contains "The Jungle Sluggard," perhaps Beebe's most popular essay. It shares with Beebe's monograph on the same subject the advantage of a single theme and focus, the incredible sloth. In general it lacks both the characteristic indirection at the start and the complexity of later development usual in a Beebe essay of this period; hence, perhaps, its popularity. But for those who relish the play of Beebe's mind over a multiplicity of facts, "The Jungle Sluggard," for all its felicity and affectionate wonder and good sense, is rather slight—a model essay by an expert whose best work probes deeper and challenges more rigorously.

"Old-Time People," the relatively brief penultimate chapter of this book, is something of a curiosity in Beebe's work, an essay written primarily from his imagination and not from his data. In it he offers comments on various species of monkey, and then his speculative overview of the way in which apes may have become men. Although this may be in part a response to the furor over evolution that produced the Scopes case, the so-called "Monkey Trial" held in the summer of 1925, it is obviously not intended as a mere legal brief or an attack on William Jennings Bryan or the state of Tennessee. The style is spare, almost grave, reminding one of the "plain style" in which early American divines were wont to clothe their thoughts. The argument itself is offered with economy, the concrete details contributing to the intended synthesis, and not (as in some other essays) complicating or obscuring it. By no means are all the questions resolved; the essay ends on a nicely tentative note, with much remaining unstated. William Beebe has said what he wishes to say, and no merit attaches to asking for what it not there; but it is reasonable, if equally fruitless, to wish that he had ventured more often into such areas, qualified as he was by decades of observing and relating.

Both of these books (as well as *High Jungle,* brought out a quarter of a century later) must be judged for what they largely are: reports from the field. Again and again Beebe identifies the scene where he is actually composing his essays—a beach, a forest, the shade of a tree at the edge of a jungle clearing. In the first chapter of *Jungle Days* Beebe mentions the Akawai Indian messenger who will take the manuscript on the first leg of its journey to "the

Editor"—which surely suggests that, except for general editing, Beebe's work was not revised, but printed as written. Given the man's learning, these are scarcely "native woodnotes wild"; they are at one level unstudied productions emerging from the complex experiences of a busy naturalist's life, swiftly perceived and recorded and sent along, and at another level, successive responses to recurrent magazine deadlines. That there are advantages to such spontaneity and urgency can be seen throughout these books, in one brilliant or arresting or delightful passage after another; that there is also value in more leisurely composition and revision is clear from the last chapter of *Jungle Days*, "The Bird of the Wine-Colored Egg." Both this and "Old-Time People" are shaped, structured, even polished, and what they may lose in zest they gain in elegance. In the books yet to come, whether concerned with land or sea, field reports would be the rule and studied work the exception.

Crusoe-like in his hunting gear, and with his wrist still bound from a flight training accident, Beebe sets out to collect specimens in Guiana in 1917. (NEW YORK ZOOLOGICAL SOCIETY PHOTO)

The crossbow allegedly used against Beebe at Sin-Ma-How is held by a Kachin camp boy. (NEW YORK ZOOLOGICAL SOCIETY PHOTO)

Mary Beebe returns from literary oblivion in a photograph taken by William himself in Sikkim in 1910—but not printed in any of his published works. (NEW YORK ZOOLOGICAL SOCIETY PHOTO)

Drojak, dubbed No-Spear-Can-Touch-Him after he killed eight men in a single encounter, is photographed with Mary in Borneo in 1910. (NEW YORK ZOOLOGICAL SOCIETY PHOTO)

Mary treats a Dyak for an eye inflammation at the expedition camp along the Mujong River in Borneo. (NEW YORK ZOOLOGICAL SOCIETY PHOTO)

A blow gun of the type commonly employed in hunting birds is used by a Dyak member of the Beebe expedition on Borneo. (NEW YORK ZOOLOGICAL SOCIETY PHOTO)

Dense jungle near Kartabo is illustrated by Paul Griswold Howes in a photo he took of himself in 1922. (PHOTO BY PAUL GRISWOLD HOWES FROM *Photographer in the Rain-Forests*)

Near Kalacoon, Beebe and G. Inness Hartley inspect two chicks of the Trumpeter *Psophia crepitans.* (PHOTO BY PAUL GRISWOLD HOWES FROM *Photographer in the Rain-Forests*)

The strange hoatzin appears in these early photographs, the adult seen on her nest and the week-old nestling shown climbing with both feet and wing claws. (PHOTO BY PAUL GRISWOLD HOWES FROM *Photographer in the Rain-Forests*)

The Arima valley near Simla, part of Trinidad's forested Northern Range, was the site of Beebe's final tropical studies. (NEW YORK ZOOLOGICAL SOCIETY PHOTO)

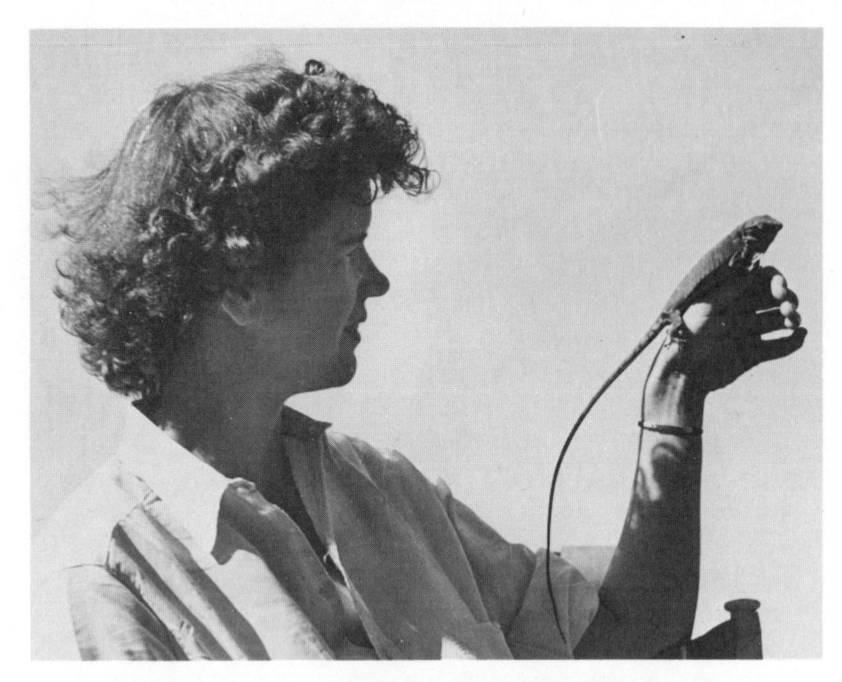

Jocelyn Crane, for thirty years Beebe's faithful colleague, took over effective direction of the Department of Tropical Research upon his retirement. (NEW YORK ZOOLOGICAL SOCIETY PHOTO)

Simla in Trinidad, the last of Beebe's research stations, was acquired in 1949, declared closed in 1971, and reopened by others in 1974. (NEW YORK ZOOLOGICAL SOCIETY PHOTO)

For helmet diving, Beebe's first fixed base was the schooner *Lieutenant*, anchored at Bizoton near Port-au-Prince, Haiti. (NEW YORK ZOOLOGICAL SOCIETY PHOTO)

:6 The Galapagos

> September 15. [1835]—This Archipelago consists of ten principal islands. . . . They are situated under the Equator, and between five and six hundred miles westward of the coast of America. They are formed of volcanic rocks. . . . I scarcely hesitate to affirm, that there must be in the whole archipelago at least two thousand craters. . . . Excepting during one short season, very little rain falls, and even then it is irregular; but the clouds generally hang low. Hence, whilst the lower part of the islands are very sterile, the upper parts, at a height of a thousand feet and upwards, possess a damp climate and a tolerably luxuriant vegetation.
>
> Charles Darwin

> Take five-and-twenty heaps of cinders dumped here and there in an outside city lot; imagine some of them magnified into mountains, and the vacant lot the sea; and you will have a fit idea of the general aspects of the Encantadas, or Enchanted Isles. A group rather of extinct volcanoes than of isles. . . . Little but reptile life is here found:—tortoises, lizards, immense spiders, snakes, and the strangest anomaly of outlandish Nature, the *aguano.* No voice, no low, no howl is heard; the chief sound of life here is a hiss.
>
> Herman Melville

So were the Galapagos Islands characterized more than a century ago by two remarkable men.[1] Darwin was twenty-six when he arrived on the *Beagle,* prepared to study the Galapagos in light of the knowledge which nearly four years as chief naturalist of the expedition had given him. Melville was twenty-two when his whaler the *Acushnet,* outward bound, stopped to take on three great land tortoises as shipboard provender. He confessed that he and his shipmates "made a merry repast from tortoise steaks and tortoise stews," despite the sympathy he had felt for these "ponderous strangers" as they lumbered in dull confusion about the deck. Then from his own experiences and the tales of others Melville wrote ten "Sketches" of those strange islands, publishing them in 1854. This was three years after the critical failure of the titanic *Moby Dick* had confirmed (as *Mardi* had suggested) the decline of this great writer into an obscurity which lasted thirty years beyond his death. Herman Melville responded as a creative writer to the exotic singularity of the Galapagos Islands, but they afforded him no more than way stations, whether on his outward voyage of discovery in 1841, or, a dozen years later, on his downward way into bitter neglect.

Charles Darwin responded as a scientist to whose powerful, probing intellect these islands disclosed a singularity of quite another kind: "The most curious fact is the perfect gradation in the size of beaks in the different species of Geospiza. . . . Seeing this gradation and diversity of structure in one small, intimately related group of birds, one might really fancy that from an original paucity of birds in this archipelago, one species had been taken and modified for different ends." And he concludes: "The distribution of the tenants of this archipelago would not be nearly so wonderful, if, for instance, one island had a mocking-thrush, and a second island some other quite distinct genus. . . . But it is the circumstance, that several of the islands possess their own species of the tortoise, mocking-thrush, finches, and numerous plants, these species having the same general habits, occupying analogous situations, and obviously filling the same place in the natural economy of the archipelago, that strikes me with wonder. It may be suspected that some of the representative species . . . may hereafter prove to be only well-marked races; but this would be of equally great interest to the philosophical naturalist."[2]

And it was as the greatest of modern "philosophical naturalists" that Darwin followed such insights to their radical and—to him—inescapable conclusions. Barring those who rejected or even fled from *The Origin of Species* (1859) and *The Descent of Man* (1871), every naturalist after Darwin necessarily shared his revolution, and none more joyfully than William Beebe. "In the wake of Charles Darwin!"—such was the spirit of Beebe's first expedition to the Galapagos archipelago aboard Harrison William's steam yacht *Noma* in the spring of 1923.

His many seasons at tropical research stations in British Guiana had afforded Beebe both the plethora of living forms and the continuity of effort needed for evolutionary investigation. Again and again his jungle studies turned up evolutionary anomalies, perplexities, even mysteries. As early as 1915, in Brazil, he had hearkened to "the question of the great black frog, *Wh-y?*" and had asked in turn whether his intense scrutiny of Amazonian jungle life had brought "the real intimacies of evolution" any closer. Often in the years following he came around to the same questions, not in despair, but always in humble awareness of the vast areas yet to be made known. And so it was nearly inevitable that William Beebe, always a seeker whose special joy was in the search, should be drawn to the strange equatorial islands where Charles Darwin's quest had so fatefully culminated.

Darwin had spent five weeks here; Beebe had available to him fewer than one hundred hours actually on the islands, but he had both Darwin's work to draw on, and that of several scientific expeditions in the decades since. Working within such narrow time limits, Beebe knew that he could not hope for comprehensive studies of the larger islands. Instead he and his party of twelve associates hoped to concentrate on the smaller ones, "mere specks on the largest charts," for their major efforts.

But it was on the second largest island, Indefatigable, that Beebe took his first walk, and he was scarcely across the beach before he was confronted by his first evolutionary puzzle. Swiftly down the sand came three birds that Darwin had called "mocking-thrush"—a near counterpart to the familiar American mockingbird, which these birds resembled in general form and plumage and even song. But they did not fly—they *ran* to inspect their visitor, stopping just a few feet away to peer up at him. Beebe was soon to discover that such locomotion was more characteristic than flying. Reporting on this experience in *Galapagos: World's End* he said: "As I look back, I remember these birds far more on the ground than in bushes. They walked and ran, they chased flies, and often leaped over obstacles without opening their wings." And he was aware that these birds, though somewhat smaller than the American species, had proportionately larger bills, longer legs, and shorter wings.

These changes in habits and structure seemed to him clearly to show adaptation to the special Galapagos environment. In seeking food, whether animal or vegetable, the island mockingbirds were likely to forage on or near the ground, often over rough lava or windy beaches where running or hopping was more advantageous than flying. The big bill was useful both for discovering food in such rigorous surroundings, and for cracking the shells of items of diet unusual for mockingbirds: crustaceans, mollusks, and even the eggs of other birds, including some of the larger sea species. Although this last habit seemed strange enough, Beebe was particularly astonished to find that in acquiring the practice of foraging along the water's edge, his Indefatigable mockers had also developed the habit of teetering occasionally in the manner of certain shore birds.

Pursuing another mockingbird enigma, Beebe took up the complex and seemingly random pattern of plumage similarities and differences found among the various island races (or species, as he preferred to call them).[3] Such irregularities, however "annoying to the taxonomist," he saw as inevitable to species cut off from the presumed parent stock when, according to the geological theory espoused by Beebe, thé present islands became separate as the early land mass subsided. However, he viewed such plumage variations as less important than the general tendency among all the island mockers toward bigger bills and legs and shorter wings, and he is properly cautious in assigning causes: "whether [changes are] initiated as internal variations or in response to external conditions we know not."

On the islet of Daphne north of Indefatigable, Beebe by chance encountered a mixed group of the birds which had intrigued Darwin even more than the mockers—the ground finches. Here were three distinct species of the genus *Geospiza,* the great-billed, the sturdy, and the sooty, readily separable by overall length and by size of beak. Such "gradation and diversity of structure" had suggested to Darwin the likelihood of an ancestral type from which these finches had descended, in the process diverging through environmental necessity. Thus, to cite the familiar example, different types of available food had resulted in different bill forms. But if such forces had once operated to produce one or another species, to Beebe they now seemed far less rigorous—witness the fact that three finch species were resident and preparing to nest in this one area of one small island. Collecting on the spot equivalent specimens of each, Beebe found no major differences in food; whether the bird had a huge bill, a sturdy bill, or a relatively slender one, each had fed on a few insects, one or two berries, and small, generally uncrushed seeds. He got the same results when he repeated the experiment with ground finches of other islands: "Birds utterly dissimilar in relative proportions of mandibles were feeding upon identical food, and food which usually showed no sign of being crushed."

Perhaps, he reasoned, the bill differences had arisen during an evolutionary period when food sources themselves were different —for instance, "the mighty beak [of the great-billed finch] was developed for coping with some source of food which had now disappeared." But more recently, perhaps with better food opportunities, there had come "relaxed environmental control"; nor could these newer conditions be reversed by predators, for here Beebe detected a scarcity of natural enemies quite unusual for any bird population. Both the accipiters, well named "bird hawks," and the falcons are absent from the Galapagos; the one resident raptore is *Buteo galapagoensis*, big-taloned and fearless, but not especially swift. The native day-flying owl, quite similar to the short-eared owl in appearance and habits, is perhaps the most significant finch predator. Except for the domestic cats that have been introduced in one or two places, no bird-hunting or nest-robbing quadruped mammal is to be found.* The great iguanas pay no attention to birds, while the two smaller Galapagos lizards and the single native snake—an innocuous species resembling the North American garter snake—Beebe found to be largely insectivorous.

He also discovered competition between one avian group and another to be unusually limited. Among birds generally classed as seed eaters, the six species of *Geospiza* (the three named earlier, plus three others less widely distributed) have the field nearly to themselves, with "no near relatives, no grosbeaks, tanagers, or buntings" to compete with, and also "no blackbirds, larks, or crows." There is, however, the native dove, and winter in the northern hemisphere may bring the migratory bobolink to the Galapagos as a temporary competitor.

Even when considered as insect eaters the finches are not especially crowded, their principal avian competitors—other than their own relatives in the sub-family Geospizinae[4]—being a cuckoo, three species of flycatcher, the yellow warbler, the purple martin and the barn swallow (another winter visitor) and especially

* Rats, dogs, and swine, similarly bestowed as blessings by the greatest predator of all, may occasionally cause inroads among the finches, but scarcely at the level of a scientific collecting party.

the mockingbird. But as Beebe notes again, this last species also lacks the usual North American rivals, there being no "brown thrashers, catbirds, wrens, creepers, bluebirds [or] thrushes" in the archipelago to put pressure on the mockingbirds. While he knows that both the finches and the mockingbirds have undergone evolutionary changes, Beebe seems to be saying that environmental necessity is no longer served by diverse bill sizes among the finches or mixed patterns of plumage among the mockers.

The phenomenon of "irregular, uncontrolled variation . . . without enough environmental stress to produce very definite distinctions" was nowhere more clear than among the seven Galapagos forms of the snake genus *Dromicus.* Not merely from island to island, but even within a single area these creatures showed "almost inextricably mingled characters" when one specimen was closely compared with another. Differences in pattern, rows of scales, scale pits, and temporals were often so marked as to suggest different species, yet so intermingled (even on a single specimen) as to defy easy taxonomic analysis. Similar if not so pronounced variations Beebe likewise found among the *Tropidurus* lizards. He gave no simple answers as to causes, knowing that genetics can work in obscure ways; but with these reptiles as with the finches and mockingbirds, he offered relaxed environmental pressure as the main factor permitting such changes to occur.

In one notable way, however, the Galapagos environment had not merely retained control but apparently had tightened it. If predators and competitors seemed to offer less than normal checks on population, a more subtle force had come into play, a kind of reproductive counterbalance. The first hint of this came to Beebe when he discovered that the island species of the purple martin laid considerably fewer eggs than its near relative on the continent. He found the same to be true of the single endemic snake, with an apparent average of two eggs against eight for related forms. The finches laid normal clutches—but about half the eggs proved to be infertile. Even the pairs of mockingbirds, for one reason or another, seemed to bring only one or two nestlings to the fledgling stage within a given breeding cycle. Thus "an adaptive diminution of numbers" had occurred, perhaps obscure in cause but clear in effect. "The absence of enemies, or the effect of some other environmental insular relaxation has apparently called forth this subtle but quick response"[5]—in simple terms, fewer offspring.

Whatever the significance of his data or the cogency of his

judgments, Beebe found in the Galapagos active challenges in an area of enduring concern, evolutionary theory. He clearly perceived that the study of evolution had not been foreclosed when Charles Darwin left the islands; the archipelago still provided a natural laboratory of a very special kind, with ramifications still to be more thoroughly examined and understood.

The tameness of Galapagos creatures, expressed in fearlessness or even friendly curiosity, remained the most striking general characteristic the expedition found. Many early visitors had noted this strange and charming trait,* and Darwin concluded his Galapagos account with a description of the tameness of the birds. At the very moment of wading ashore on Indefatigable, Beebe's party attracted a small duck, which flew in and paddled alongside and then waddled in their midst right up the beach, quacking and looking up curiously into their faces. Later the three mockingbirds approached, one singing as it ran, and when Beebe moved inland through the brush, they often alighted at arm's length to look him in the eye. Then a Galapagos short-eared owl flew up and perched on Beebe's hat, causing the mockers to scold suspiciously, but neither to cower nor to flee. Evidently the mockingbirds' toleration of *Homo sapiens*, the fiercest predator in the planet's history, extended also to this modest predator *Asio galapagoensis*.

A similar example of benign indifference occurred on the islet of Eden, where Beebe's party saw a scarlet rock crab walk the entire length of a three-foot marine iguana, meanwhile picking off parasites to eat. In like manner the small lizards of the islands ran heedlessly over the bodies of their giant (and vegetarian) cousins, and these great reptiles were equally unconcerned about the near presence of creatures they had perhaps never encountered in their lives, vastly larger than crabs or *Tropidurus* lizards, and with two legs instead of four or eight.

At one point Beebe made an experiment to find out whether the marine iguana could be taught the kind of fear reaction denied it by a long history of isolation. His expedition was charged with the task of bringing back to New York numerous iguanas for display (though they appear to refuse all food in captivity) and the method of capture was simply a noose on a pole. One day Beebe

* As Beebe says, "Ever since the time of the very first visitors to the archipelago this has been remarked, and most human observers have celebrated this fearlessness by knocking as many as possible on the head."

caught a fair-sized iguana in his noose, jerked it into the air and swung it roughly around, and then released it. Although it ran off a few feet, this iguana allowed itself to be recaptured half a dozen times and subjected to the same rude aerial experience. Yet from a series of events which to the common wild creature would have been terrifying, it developed no fear at all. At no time, in fact, did any marine iguana show fright or truculence or anger, and the only resistance came when Beebe tried to pull one or another from rocky crevices. By approaching sea-lion fashion, Beebe found that he could get close enough to stroke the creature's scaly skin, and capture it without effort. It scarcely needs said that sea lions and iguanas shared quite amicably the same seaside rocks, the latter occasionally crawling unremarked over the bodies of their ponderous neighbors.

Sea lions were equally fearless toward human intruders, but more active in their responses. Initially, indeed, Beebe was given good reason to believe that these seals were downright pugnacious. On the morning of his second day in the archipelago, he landed on Eden and was almost immediately assailed by an angry male sea lion. The great creature rose with a roar from the water's edge, evidently detecting in Beebe some sort of rival, however misshapen and attenuated. He surged up again and again just a few feet offshore, growing bolder as Beebe, crouched down and craftily retreating, encouraged a final attack that stranded the beast on the pebbled beach. Instantly Beebe attacked in turn, grabbing the sea lion's hind flippers and sending the creature into a panicky, floundering retreat amid flying stones and a froth of sea water. Perhaps, as Thoreau felt toward the fox he nearly ran down in deep snow, Beebe had taught this rogue male a lesson—but the reverse was also true. Later encounters with male sea lions, none of them belligerent toward their human visitors, reminded Beebe of the risk the field naturalist takes when he generalizes from single instances.

Only a day later, on one of the islets of the Guy Fawkes group, Beebe was again confronted by a sea lion, although this time it was not a choleric old male but a playful youngster. In all innocence it gave Beebe quite a turn by swimming up under water and nuzzling at his hands as he stood waist deep and tugged at a mollusk on the bottom. Beebe's leap of fright in turn frightened the little seal, which surfaced and let out a wail that brought its mother and two other youngsters to the scene posthaste. Perceiving, however, nothing more fearsome than a dripping scientist, the four watched

Beebe return to his mollusk, swimming around a little distance off and now and then diving to repeat their inquisitive nuzzling.

There were many other sea lions on this islet, often females with nursing young. Many of the adults had fallen so soundly asleep that they were not roused even when Beebe walked up and shoved and slapped their sides; but the pups were usually wide awake and frolicsome. When Beebe found a natural throne among the rocks and sat down to rest, four baby seals just a few feet away peered at him with guileless interest, but were soon diverted by roughhouse play and then by the arrival along shore of Miss Ruth Rose, the expedition's historian. Here was a more promising new companion for a romp, so at their best caterpillar speed they deserted Beebe forthwith. He noticed they played insouciantly enough in sheltered pools, but did not venture beyond the outer rocks into blue water—for there Beebe could see the sharks tirelessly cruising and waiting.

So the struggle for existence had not ceased here on the Galapagos, though some of its common manifestations may not have been evident. A member of the expedition once saw a huge tiger shark suddenly rise and seize a young seal which had swum in confusion toward deep water, and others saw the smaller birds flee in fear from the local hawk and diurnal owl. "Relaxation of environmental pressure" was after all only partial; the great, indeed nearly unique difference these islands could claim was the virtual absence of the human being from their evolutionary history. Some archaeological evidence suggests that aboriginal settlements may have existed here, but in this remote and inhospitable setting they appear not to have endured for long, and no human inhabitants were discovered by the first Europeans to sight these islands.[6] Incursions by Europeans have been sporadic and usually brief; as Beebe says, "Generations of [Galapagos] creatures came and went without ever seeing a human being." Not so the birds from far away; in his first hours ashore on Indefatigable Beebe had seen and put quickly to flight the Hudsonian curlew and the black-necked stilt, migrants or recent arrivals with the common ancestral fears. The next day on Eden Beebe saw a pair of Galapagos oystercatchers with great coral bills fly down to a waterside ledge, not at all intimidated by the novel presence of a camera and tripod nearby—quite otherwise, as they immediately showed by running between the tripod's legs. But a turnstone, alighting a moment later, reacted very differently, glancing nervously at the apparatus and abruptly fleeing as quickly as Beebe made a slight movement—"no native this, but a migrant from the land of men and fear."

Over the centuries since men began coming from the paleface lands, scarcely a creature on the Galapagos has failed to suffer from its own tameness and trust. Dwellers along the forbidding lava shores or the great cliffs, the seals and marine iguanas and sea birds, have survived in places by their very inaccessibility, and by having refuges in the waters of the sea or in the air above it. But the great land dwellers, unsuspicious, inoffensive, and generally slow of foot, have had no comparable sanctuary. True, the land iguana digs a burrow into which it can retreat, and when pressed this great lizard will flee, and when seized, will try to bite its assailant. Yet Beebe found this creature neither wary nor elusive. The first he saw, a large specimen clawing at the luggage of the people who had come to explore the island of South Seymour, he quietly crept up on and caught with no difficulty. Shortly he noticed other members of the party (including Ruth Rose) returning, each bearing his or her own captive iguana. By lucky chance, no one had allowed fingers or toes to come within range of efficient reptilian jaws. In all, eighteen of these creatures, some nearly four feet long, were taken alive to the New York Zoological Park.

The usual fate of the land iguana at the hands of man was to be killed for food or even random sport, or to be driven out by man's attempts at farming, or fall victim to the destructive mammals he has introduced, for example dogs and pigs. When Beebe came to the Galapagos in 1923, iguana populations on some islands had already been nearly or wholly exterminated; so if he was less than sanguine about the future of the species, he was unhappily right. Nothing had yet been done to protect these unique creatures, nor was effective action in prospect. Ecuador, the parent country, tried setting aside portions of the islands as sanctuaries, but enforcement was ineffectual or lacking. The first serious effort to protect the unique Galapagos environment came with the creation of an international body called the Charles Darwin Foundation for the Galapagos Islands in 1959, and the building of a research station under foundation auspices at Academy Bay, Indefatigable, in 1961. In that year the land iguana was reported in need of immediate protection, and special areas were suggested as refuges.[7] Lacking these, the creature faced a doubtful future.

For the other great land dweller, the giant Galapagos tortoise, even worse prospects were reported: "great danger of extinction because of the relentless and uncontrolled hunting." Such hunting had a long and melancholy history, going back at least as far as the seventeenth century, when the first seafarers began raiding the tortoise colonies for meat and oil. Something of a climax came during the great age of whaling celebrated by Herman Melville. The Encantadas, so named not for their delights but for their strangely malign winds and currents, their desolation, perhaps even for the "fleetingness and unreality" of their location in the early days of sail, at least provided whalemen with a change of diet. Available logbooks show that in almost two hundred visits over a forty-year period American whalers took more than 13,000 Galapagos tortoises.[8] This figure obviously does not include whalers of other nations, nor American whalers whose logs give no data or have not survived—for example, Melville's *Acushnet* out of Fairhaven, Massachusetts. If Melville is telling the literal truth when he says that only three tortoises were taken aboard on the ship's first visit, then its toll was modest indeed—about one-twentieth of the average.

Though whale hunting under sail declined sharply toward the end of the century, that did not mean respite for the tortoises. Raiding parties of another kind, in this case principally from Ecuador, took the place of the whalemen, slaughtering the great reptiles mainly for their yield of oil. Added to the devastation of sailors and oil hunters were the random depredations of colonists and their introduced mammals, and of scientific and less-than-scientific collectors—all in all, a formidable threat for any species to face, and in particular one both conspicuous and defenseless. By 1906 a California Academy of Sciences expedition found that the tortoise was extinct on two of the major islands, nearly so on another, rare to very rare on five, relatively numerous on two, and abundant at only one location on Albemarle.[9] And of course further invasions by oil hunters were still to come.

Knowing all this, William Beebe was scarcely astonished when his party found only one tortoise, and that with difficulty. In 1906 the population on the island he chose, Duncan, had been given as "fairly abundant," but evidently times had changed for the worse. When at last the expedition's chief fisherman stumbled upon a living specimen of *Testudo ephippium* in a thornbush thicket deep in the island's central crater, his "yell . . . reached all over Duncan."

Scarcely a monster, however; the heaviest member of the genus reaches nearly five hundred pounds, whereas this tortoise weighed only a little over forty.

As the members of the party examined their prize, they could all too readily understand how grievously vulnerable was this antique creature. Its only defense was to withdraw into its shell, protecting its forward parts with massively scaled wrists. It had no offense, nor could it escape by running. And two and a half centuries of persecution had not yet implanted a salutary fear of human beings; once captured, Beebe's tortoise soon failed to defend itself even by withdrawal.

The history of this reptile in the hands of the expedition was brief. The return to the *Noma* took many hours, ending long after dark. A day or two later Beebe got motion pictures of the tortoise orienting itself on a strange island (Indefatigable) and invariably setting out toward the island's center. Other pictures were taken to show how remarkably proficient the creature was in climbing slopes which seemed impossibly steep and rough. The last of the experiments was perhaps its undoing: "we tossed the Duncan individual overboard and took hundreds of feet of moving picture film" to demonstrate the important fact that it could swim strongly and with an understanding of its situation. First the tortoise swam to Beebe's rowboat and tried to climb aboard, but found it too high; then it swam to the yacht's companionway. Such aptitude—either ignored or denied by other observers—suggests that the early colonizing of the Galapagos may have occurred by sea, with the ancestral form of the tortoise coming from the mainland and then spreading among the islands to evolve into the present giant races of a single genus. Holding, however, to the land-bridge and subsidence theory, Beebe denied this conclusion, and felt substantiated when the tortoise died several days later from bowel and lung congestion apparently caused by too much sea water.

In the area of Tagus Cove on the largest island, Albemarle, the expedition came upon the two most distinctive Galapagos sea birds, the penguin and the flightless cormorant. No great mystery attends the arrival of either bird's progenitors; the original cormorants presumably flew to the Galapagos (perhaps resting and feeding en route) and the original penguins swam northward along the cold currents from the Antarctic, eventually taking up residence here on the very equator, many hundreds of miles from their nearest relatives in Peru. As in Antarctica, no land predator threatened the penguins, and in the absence of such predation the cormorant evolved to its present flightless state. The wings diminished to half their normal length but did not evolve into swimming flippers, as had occurred eons ago with the penguins. The cormorant's great sturdy feet were made to serve for travel ashore at nesting time, and they provided, as Beebe in his diving helmet was to discover two years later, "a pair of most efficient propellers" for catching fish.[10]

Thus by the time the foremost terrestrial predator arrived on the islands, neither bird could escape by flying, and both could be caught easily because they had never learned to fear this strange new creature. The little penguins stood amiably waiting to be netted, and the first cormorant Beebe approached steadfastly brooded her single egg until he reached to pick her up—but then she dodged and thrice raked his hand with her curved bill tip before she could be subdued. Shortly he secured another, "getting full measure of wounds from her as well." Aboard the *Noma,* according to Ruth Rose, one of the cormorants "promptly laid us an egg like a well-trained fowl, and we transported them to New York without mishap. The three penguins that were captured made irresistible pets, and were so tame that it was almost impossible to keep them at arm's length long enough to take their photographs."

Collecting Galapagos flora and fauna, whether for the New York Zoological Park or the American Museum of Natural History, was an assigned task. Adventuring was not, though it might at any moment produce a new specimen or a novel fact. Dismissing the idea of trying to reach the interior of the major islands, the expedition used its limited time in exploring smaller ones. As always, Beebe wished to observe closely, even intimately where he could, and in so doing capture a sense of place and mood and natural scene. This he could accomplish on such islets as Eden and Guy Fawkes, and this he strove to do on the other small islands he chose to explore.

After capturing their first land iguanas on South Seymour, the expedition people climbed the cliffs beyond the beach area to discover a high plateau, an area of grass and cactus and low trees stretching like the African veldt across the center of the island. This itself was singular enough, since volcanic cones and craters dominate so much Galapagos scenery—and then the party saw a small herd of antelope-like animals with wide, spiraling horns, to complete the veldt illusion. They were in fact goats gone partially wild since being introduced years before. They had prospered and grown sleek, perhaps because there were pools of drinkable water here—scarcely the case on most of the islands; and yet Beebe found many puzzling skeletons of goats which had died apparently in the prime of life, from causes he could not readily perceive. Later along the Seymour beach he came upon half a dozen skeletons of quite a different mammal: false killer whales, a little known species never recorded for the Galapagos, nor anywhere closer than Greenland or Tasmania. Perhaps they had died of stranding, since all six skeletons lay above the regular high tide mark.

As compared to areas near the beach, the Seymour veldt had many more land iguanas, some in groups that crowded each other for the slight shade cast by bushes and cactus plants. It was interesting to see that these big lizards occasionally fed in trees, and might leap or simply tumble to the ground when human beings approached. Their usual response at ground level was to move slowly off, perhaps toward burrows, but even if they ran they could be easily overtaken. They also had rather witless ways of hiding, pushing head and foreparts into a crevice or the shelter of a bush and placidly waiting, confident of concealment. Once captured, they wasted little time sulking, and, unlike their marine counterparts, these land iguanas accepted food eagerly and subsequently thrived in the confinement of the New York zoo.

West of Seymour lay another islet that promised good exploring, Daphne Major. It was less than a mile across and consisted mainly of a truncated peak enclosing an ancient crater. However, beyond such meager facts little was known, since very few men had actually visited the place. Lack of a beach may have been one reason; the members of the party were obliged to land on a jutting ledge, stepping quickly from their boat when the ocean swell lifted it to the right level. From there the party spread out, each seeking his own way up to the crater's rim.

Beebe soon came upon a breeding pair of black finches which foreshadowed the anomalies he was to find among a colony of such birds near the top of the crater. The single young, still being fed by both parents, had already reached the fledgling stage; nevertheless the building of the nest continued, and inside was a single fresh egg. Dissected later, the female was found to be carrying another fully formed egg, with a third in a late stage of development. The male of the pair was in mature breeding condition but partly immature plumage. Furthermore, the food brought to the fledgling by these sturdy-billed finches consisted of insects as well as the expected seeds. Such oddities were duplicated and even exaggerated by the colony of three finch species Beebe encountered half an hour later, leading him toward the theoretical conclusions noted previously.

At the crater top Beebe paused to revel in the view before him. In the day's bright sun the sea lay calm and intensely blue, with islands great and small rising from its mighty deeps: James and its satellite Jervis, huge Albemarle and little Duncan, tiny Eden and Guy Fawkes, all in the sweeping arc of the western horizon; just to the south, great Indefatigable; to the east the two Seymours; and Daphne Minor northeastward.[11] When at length the crater drew him, Beebe chose a relatively easy path along its vast curved slope and made his way down. Here he found what the reports of two earlier visitors had led him to hope for—the entire level bottom of the crater taken up by nesting birds of one species, the blue-footed booby. The goose-sized adults were dusky above and white below; most of the young were downy white. The nests—mere firm spots in the sand—numbered about 400, the birds themselves many more, and their calling was so clamorous that normal conversation was impossible.

The most interesting of Beebe's observations here concerned the highly special setting afforded by a crater as a booby nursery.* The steep crater walls, hundreds of feet high, enclose the young booby from the moment of its hatching to the moment of escape, months later, on its five-foot wings. In the absence of terrestrial natural enemies, such walls are scarcely meaningful as protective barriers; but at a crucial moment they afford a stern test of individual survival. Parental feeding eventually ends, and when the time comes the fledged offspring must find its own food in the sea or die. Of course it has never seen the sea, but its instinct for flight can take it there—if it can surmount its nursery walls. There is nothing so arbitrary here as a marauding predator's random choice among the eggs and chicks of a mainland rookery; this is a test of each bird alone against implacable gravitation, a trial by the ordeal of flight. Most survive and reach the sea, but some do not. Of the bodies found within the crater, the larger number were of birds just fully fledged, those that had perished on the threshold of ma-

* Beebe, who once used the phrase "tunaville trolley" for the hand carts in a fish cannery, here merits boundless praise for resisting the term "booby hatchery."

turity. In this rigorous selection process Beebe saw the survival of the fittest working to benefit the species in a way far exceeding in value the accidents of common predation.

Yet there is another kind of survival having nothing to do with natural fitness, but only with accessibility. As they pulled away from Daphne's shores, the expedition people noticed that for all the abundant life hidden within the crater, scarcely a bird was to be seen flying in or out. The bleak moral of the story comes out in Beebe's concluding sentence: "May the little island long keep her secret from devastating mankind—a single boatload of which could exterminate the whole colony in a few hours."

About sixty miles north-northeast of Indefatigable lies Tower Island, small and relatively low, the last and in some ways the most remarkable of the islands which the party explored. One feature was so inexplicable as to be truly mysterious—the presence of a deep sheltered bay, over a mile across, which apparently had not been mentioned by any visitor or shown on any extant map in two and a half centuries. True, except at a certain angle the entrance is fairly well hidden from vessels lying offshore, but when Beebe and members of his group set off in boats from the *Noma* and made a circuit of the island they found it readily enough, and, venturing in, they discovered this wide inlet of the sea and named it Darwin Bay. The date was April 18, 1923.

Whether the bay would serve as an anchorage for the *Noma* was still in doubt. When the expedition returned nine days later, careful soundings were taken showing an entrance nowhere less than twenty-seven feet deep, which provided water to spare under the yacht's keel. She steamed cautiously in and anchored at seventeen fathoms. By then the party in boats had found a beautiful spot for a base camp, a gently sloping white sand beach with a tidal lagoon beyond.

The shores of this splendid bay offered many surprises. In one small cove Beebe found the beach piled with pieces of coral, something of an oddity in this lava land, and betraying the presence of living coral in the clear shallows nearby. Beyond lay a mangrove thicket crowned with many nests of boobies; further along was a small secluded beach where sea lions lay basking in happy torpor, belly up; and elsewhere the shore rocks were crowded with marine iguanas of half the common size or even less. Here and there were estuaries to be skirted, or tide pools each with its own pulsing life to be watched and wondered at. Then by one rocky stretch of shore

Beebe found himself suddenly under attack by a two-foot moray eel which strangely leaped up at him from the shallow water, swiftly and fiercely but without effect, and not once but several times. Here was an attack even more startling than that of the rogue sea lion, and like that strange onset, never to be repeated, though Beebe in years to come was to venture among morays many times bigger than this oddly infuriated small assailant. It almost seemed that the benignity of the Galapagos environment had produced two bizarre deviates, creatures which had fully crossed the spectrum from gentle tolerance to rage.

From the time of their first approach to Tower the expedition people had been aware of the wonderful plethora of birds. As the *Noma* drew near she acquired an escort of frigate-birds and boobies in the hundreds, with shearwaters in flocks alongside. Ashore it was much the same, the sky appearing as "a closely-woven mesh of flying forms, a frigate-bird warp with a woof of boobies and gulls." Tower Island, fairly flat and brush-covered, seemed a single great nesting ground and sanctuary for birds of many species, from the great man-o'-war with his eight-foot black wings to the mockingbird and the dove.

The workable yet competitive relationships among the birds are vividly suggested by Beebe's tale of stolen nest twigs. Having found a kind of natural armchair in the midst of a dense thicket of two-foot *Mentzelia* bushes, Beebe sat down to observe the ways of the man-o'-war or frigate-birds, the boobies, the doves, the gulls, and the mockingbirds, all of which were nesting or moving about in the immediate area. One frigate-bird home—a fairly ramshackled platform of sticks—was within easy arm's reach, and upon it the male bird sat paying Beebe not the least attention. Even stroking the bird's back caused only a slight turning aside, but when Beebe pulled one wing upward by its tip the bird merely opened the other and floated gently up and into the wind. Then, however, came thievery. Within a brief time half a dozen other frigate-birds swooped down over Beebe's head and stole nest sticks—scarcely a neighborly proceeding, but normal enough in the man-o'-war tribe, and in others as well.

Presently such theft was extended beyond the tribe into aerial robbery. A brown booby, also engaged in nest building, was discovered by keen frigate-bird eyes flying homeward with a three-inch twig in its beak. Almost instantly the unfortunate bird was beset by a flock of frigates, and, outmatched by these masters of flight, was forced to drop its small prize. Beebe was interested to discover that his own bird had caught the extorted bit of wood and was bringing it back to the female, which meanwhile had arrived at the nest. After ceremonies of mutual congratulation, the twig was pushed into the nest structure and fell without delay to the ground. Some minutes later a mockingbird spied the twig, appropriated it for himself, and made it part of his own nest—thus for the moment at least bringing its peregrinations to an end.

The usual booty of the frigate-birds was made up of more important objects than twigs. The food of boobies is mainly fish, which they catch in swift dives that often carry them many feet below the surface. Fish make up a good part of the diet of frigate-birds also, but these birds do not dive, and indeed might perish if they tried, given their inability to swim and to rise into the air from the surface of the sea. On Tower Island as elsewhere, frigate-birds let their neighbors the boobies do the fishing, then relieved them of their catch on the way home. Beebe saw this kind of mid-air piracy enacted many times, and with much the same result in each instance—the beleaguered booby was forced by grim frigate-bird harassment to disgorge most of its catch, and was "lucky if it got through with a single fish."

These events, in themselves unremarkable, were only a few among very many which might be noted and correlated by an ornithologist in the Galapagos. Beebe's observations on Tower Island were numerous and varied, but they did not constitute an integrated whole, as he himself was aware. For example, he mentions the fact that although many tropical regions have nesting areas shared by frigate-birds and boobies, the interrelationship of the two species has never been intensively studied. Nor was he prepared to do so himself; time pressed, and his several interests and responsibilities divided up whatever hours were available. Next time perhaps; this was after all his first seaborne expedition, and he hoped there would be others. As the *Noma* steamed away toward the continent late in the evening of April 29, Beebe prophesied "other golden days to come in these islands of enchantment"—and two years later he was back.

The new expedition had been set in motion on the same day the *Noma* returned to New York, May 16, 1923. The industrialist and yachting enthusiast Henry D. Whiton offered the steam yacht *Arcturus*, if Beebe could find somebody to supply the coal. Although Whiton's fellow member of the Zoological Society's Board of Managers, Harrison Williams, had sponsored the trip just concluded, this generous utilities magnate again volunteered his support, ultimately contributing by far the largest amount of money. That the total amount was fairly formidable is suggested by the names of some of the other donors: Vincent Astor, head of the noted family; Clarence Dillon, banker; Marshall Field, mercantile heir and corporate director; and Junius Morgan, son of J. Pierpont Morgan and partner in the family banking firm.

Much of the money went to renovate the yacht, a wooden vessel of 2,400 tons displacement launched in 1919. She was to be transformed from a pleasure craft to an expeditionary mother ship, with laboratories and darkrooms and new storage space for specimens and oceanographic apparatus. The scientific staff was to be somewhat larger than that aboard the *Noma,* and the scientific equipment more elaborate; and whereas the *Noma* was at sea only two and one half months, the *Arcturus* was expected to log twice that and more.

So the new expedition was in general a more ambitious undertaking; and it was also more coherent in purpose, setting out with the intention at least of concentrating on oceanography. Some of the titles for the staff indicated this new bias: Associate in Diatoms and Crustacea, Assistant in Larval Fish, Assistant in Fish Problems, Assistant in Macroplankton. William Beebe was shifting his emphasis still further away from the land and the life of mountains and plains and jungles, and more toward the sea and the numberless creatures living there.

The avowed and specific aim of this expedition was to study both the Saragasso Sea south and east of Bermuda and the Humboldt Current which moves up the Pacific coast of South America to reach the Galapagos. The *Noma* on her third day out had passed through sargassum weed which the party scooped up and studied, and evidence of the Humboldt Current had appeared both in the abundance of aquatic life and the coolness of the water. Now these chance data were to be augmented by systematic observation, collection, and analysis. Such was the intention; such was not the outcome in either case. The *Arcturus* sailed from Brooklyn on February 10, 1925, stopped at Newport News for coal on the 13th, and set out for Bermuda the next day. Not until the 21st did the amount of floating weed suggest the real Sargasso Sea, but this impression faded as the weed quickly thinned out and was never seen again in patches appropriately wide and dense. Beebe could only conclude that winter storms had scattered the Sea too widely for useful study. After proceeding halfway to Africa, making interesting catches with nets and dredges but never encountering real masses of weed, Beebe gave up the effort and ordered the yacht to head for the West Indies.

Off the islands of Saint Martin and Saba, hauls were so rich and varied that Beebe wished for a year of study instead of two days here on the rim of the blue Caribbean. But the course for Panama was set March 16 and on the 21st the yacht dropped anchor near Fort Sherman in the Canal Zone. Five weeks at sea amid much foul weather had already taken a toll—both the ice machine and the vital circulation pump needed repairs. Always ready to take advantage of such delay, Beebe went off observing and collecting birds and snakes and other creatures at a favorite place called Devil's Hole. The last few members of the expedition joined here in Panama; then on March 28 the *Arcturus* traversed the canal and set off for the Humboldt Current and the Encantadas.

She found the islands easily enough, but never the current. This failure to Beebe was "inexplicable"—and yet many years later, when the same anomaly had recurred and had been investigated under the name "El Niño," this sporadic warm-water incursion was still not fully understood. The consequences for Beebe and his party were various: delightful swimming, unexpected rains, far more greenery than they had reason to anticipate, and of course, with the current playing truant, opportunities to study other aspects of these enthralling islands. There are times when the log reads more like a vacation diary than a report of scientific activity: "Everyone ashore. . . . The place is as wonderful as ever. . . . Most of us swimming for hours. . . . Everyone groaning with sunburn to-night."

For the better part of three days the *Arcturus* lay anchored in Seymour Bay; then at midnight on April 6 she departed northward for Tower Island, dropping anchor in Darwin Bay late the next morning. Here as on Seymour and Indefatigable, the party went ashore to renew old acquaintanceships with favorite creatures and to continue observing, exploring, and collecting. More systematic efforts were begun April 10, with the scientific staff divided into separate groups to concentrate on special tasks: "shore collecting, plankton collecting, fishing, identification and dissection of fish caught, painting, photographing, mapping and sounding the bay, and diving in the helmet." Given all this, one is not inclined to dispute Beebe's word "tremendous" for the activity shown that day; and one notes furthermore a new kind of aquatic work, helmet diving—for Beebe a means of study destined to become increasingly engrossing.

Long before dawn the following day, a great natural phenomenon intervened to change for a time the course of the whole expedition. Shortly after midnight the officer on watch on the bridge of the *Arcturus* saw a strangely flaring, reddish glow on the horizon to the west, and soon roused the ship. Somewhere among these dark islands, themselves born of molten lava, new volcanic fires had burst forth. Thirty miles to the west lay Bindloe, and twice as far to the southwest, James; but the charts placed the center of the red glow on the northern part of Albemarle, nearly 100 miles away.

By morning the sea had risen so much that even in the shelter of Darwin Bay the *Arcturus* began to drag her anchor. Without waiting to take aboard the small boats pulled high on the beach by the base camp, the yacht cleared the bay and set out for Albemarle. Oceanographic work done en route could scarcely claim the party's fullest attention. As the day passed the ocean grew calm. With darkness the volcano's fires shone again, and few people slept the night through. At dawn, as the red glows paled in the light and the great billows of smoke appeared against the sky, the *Arcturus* lay only a few miles offshore.

With twelve-power binoculars Beebe scanned the zones of fire, discovering the main upwelling of lava to be along a saddle between two of Albemarle's peaks. At the same time, altogether characteristically, he was searching out the route he would take to reach one of the largest active craters. William Beebe had never before seen a volcano come to birth; now he would embrace the experience, he would know it fully and directly, and not from afar. Soon a boat was brought around and he and John Tee-Van dropped into it and made for shore. At length they found a small cove where they had no trouble landing, and set off toward the chosen crater after telling the boatman to return in three or four hours. Both men wore high leather boots and light clothing; they had one canteen of water between them.

Perhaps they were deceived by the easy path they found at the start—an older lava flow hardened almost to black glass, and smooth underfoot. But then came "hellish rock froth" over which they were forced to struggle, never sure from one step to the next whether the surface would sustain their weight, or would crash through in lacerating shards. This was not an ash-fall such as the one that long ago covered Pompeii, nor a slide like the one which engulfed and preserved Herculaneum; this was the remainder of a seething molten river which had thrust itself into strange currents and waves and froth as it hardened and cooled. But even "cooled" seemed a mocking term to the men as they toiled along; the sun's heat pressed down on them from above, and then, as the lava absorbed the heat, it billowed and radiated upon them from below.

More than once Beebe was forced to change his plan, choosing smaller and nearer craters as his goal. Exhaustion was too close to permit a greater risk. The air near ground level seemed to be growing noxious; hence Tee-Van and Beebe were even denied the partial respite of squatting on their heels to rest.

It was almost by accident that Beebe at last stumbled upon a small crater which showed him the things he had ventured so far and painfully to see—gas puffing from vents, molten lava glowing in holes. Visible gas drove him toward an area of the crater seemingly free of such fumes; but then he realized that some other gas, invisible but deadly, surrounded him. He was seized with nausea; his sight began to dim. Before his strength wholly failed he staggered from the crater, slid recklessly down the slag-like outer slope, and with John Tee-Van started on the tortuous journey back, "too exhausted to do more than choose whatever way seemed least terrible." When at last they found the beach their drinking water was long gone. They lay in the cool salt water, recovering from cramps and regaining the power of speech. It had not been three or four hours, but five.

Copious imbibing aboard the *Arcturus* rescued them from dehydration; Beebe had eight glasses of water and a bottle of beer. Then, scanning the fiery lava fields once more, he could laugh at his own temerity.

For another two days the yacht lay off Albemarle. The volcanic fires did not diminish, and even when the expedition returned to Tower Island the volcano's vast dome of cloud could still be seen on the distant horizon. As far as anyone knew, the only human witnesses to this natural wonder had been the fifty-six people on board this one vessel. Then, like the *Noma* two years before, the *Arcturus* was forced to run for Panama for supplies and repairs, with pauses for oceanographic work en route, and a ten-day stop at the "pirate island" of Cocos on the return trip. It was nearly nine weeks before they saw the eruption again, noting at once how much things had changed. In the interim the fires had spread from the saddle between the peaks downward toward the sea, flowing in unseen molten veins beneath the ground. As the yacht arrived off-shore these veins were just beginning to burst over the top of the shore cliffs and spill in scarlet cascades into the Pacific. Beebe estimated the height of the cliff at one point at about 100 feet, and timed the speed of the plunging lava at two seconds, top to bottom. And this was only one molten river out of nine or ten along the seething shore.

Approaching as close as they dared in a stiff landward wind, they watched in awe the apocalyptic joining of lava bright as blood and the waters of the sea. Great spurting coils of steam and sulfurous, boiling froth billowed upward and away on the wind. The ocean swells that rolled in upon the molten shore rose up and broke green and white and vanished into vapor. From beneath the cauldron's surface came roaring bursts of gray gas and fragments red and black, born of lava exploding as it was quenched. In such mighty heat the water changed to a clear pale green, sharply marked from the deep blue offshore. At one point a delta of this green water flowed a thousand feet to sea, and the *Arcturus* sailed directly across it. Just athwart the line dividing the green and blue, a thermometer dropped from the bow read seventy-eight degrees, and from the stern, ninety-nine degrees.

From this heated water many fish fled in panic, but some not fast enough. Especially in the seething waters of the shore, fish died or were disabled, and static organisms perished with no chance to escape. To the grim banquet so spread the scavengers hastened, scores of frigate-birds and petrels and shearwaters, with a scattering of boobies and brown pelicans. Some flew and dived so close to shore that they were obscured in the rolling clouds of steam, and a few died from the heat or the poisonous fumes and were seen floating. The most awesome death was that of a sea lion which, perhaps in the pursuit of prey, blundered into the scalding coastal waters, leaped blindly from the surface several times and vanished forever.

Those aboard the *Arcturus,* again the only human audience to this elemental play of forces, watched through the day and into the darkness; and as they steamed away, nine scarlet streams of lava were still cascading from the black unseen cliffs into the night sea.

The next day the steering gear failed suddenly and completely, and until an emergency system could be rigged the ship drifted with the breeze. Suppose, Beebe mused, this had happened a thousand feet from the fiery coast just one day earlier, the wind strongly onshore and the ship itself built of wood. The *Arcturus* would doubtless have added, he coolly surmised, "a new odor and a few flying sparks" to the general pyrotechnics, "and after that the steam and gas would have continued as usual, and the lava flowed uninterruptedly."

A month and a half of his voyage still lay ahead, but on this fifteenth day of June, 1925, Beebe left the Galapagos for the last time. The immediate experiences were over, reduced now to notes, drawings, photographs, statistics, specimens, recollections. From these would come Beebe's next book, *The Arcturus Adventure*—published less than a year later—and the technical papers, the analyses, the reflections and theoretical conclusions. In the quarter of a century he had been working as a biologist, Beebe had encountered many of the specialized disciplines involved in the study of the natural world, from astronomy and meteorology and paleontology and geology. Characteristically, he had found things to interest him in each, and had followed his own bent in arriving at conclusions. He was not a specialist, and said as much; he was a naturalist of long experience and deep engagement, capable of weighing and judging among the findings of the specialists, and altogether willing to do so.

In *Galapagos: World's End,* for example, Beebe had stated his preference among the divergent theories concerning Galapagos origins. He believed that the islands had once been a single land mass, and hence were not the result of separate volcanic eruptions. Instead the individual islands had resulted from subsidence of this mass, with only uplands and peaks remaining above water. In the process various plant and animal species became isolated and therefore subject to further variation.

But how did plants, insects, spiders, reptiles, birds, even crustaceans and fish, originally get here? Darwin believed in the oceanic origin of the Galapagos, thus denying the notion of a former land bridge to the continent. Beebe was willing to agree—but on the basis of the published work of others and of observations made during his first expedition, he at length acquired "a strong belief" in the land bridge idea. In support he cited prior botanical studies and his work in ornithology and the marine sciences. Comparing

Galapagos species with their mainland counterparts, he discovered far more affinities with Central American forms than with those of South America; hence he assumed that the land connection had existed not with Ecuador, about 500 miles to the east, but with Coasta Rica, 650 miles to the northeast along what is now called the Cocos Ridge. Presumably that ridge, visible today at Cocos Island, had once joined the present Galapagos Island region with the continent.

How otherwise, asked Beebe, could Galapagos shore fishes—nearly all of them with counterparts along the coast of Central America—have reached these distant waters? How could the Grapsus crab, a Central American shore dweller and indifferent swimmer, have crossed hundreds of miles of deep sea? The same general question could be asked about the arrival of terrestrial species, from plants to reptiles; and Beebe could find no plausible answer except in terms of a former land connection.

However, he confessed his willingness to relinquish his theory "at the first hint of better proof on the opposite side"—and two years later, as the result of studies aboard the *Arcturus,* he provided his own refutations. No such purpose was intended, and the occasion had nothing directly to do with geological or evolutionary theory; nevertheless "better proof" arrived in abundance. Beebe had decided to establish a sea station for dredging, sounding, netting, and so forth, about sixty miles south of Cocos Island. This was nothing new for the voyage—in fact, this was the seventy-fourth station in three months of ocean travel. But it was unique in duration; the *Arcturus* hove to or slowly circled in this spot from May 25 to June 3. So fixed a situation became in Beebe's mind a kind of island, and he found himself peopling this imaginary place with whatever organisms could be discovered arriving by air or water.

First came the birds, not merely flying about the *Arcturus* but alighting on her and even (in the case of an entrancing fairy tern and two infuriated boobies) invading Beebe's cabin during a tumultuous night of wind and rain. Two other arrivals were a Galapagos gull, more than 300 miles from its native shores, and a yellow warbler, possibly from an equal distance, but more probably from Cocos. In its majestically detached way, a frigate-bird flew past one day without even pausing to rest. Among the true pelagic birds was one complete surprise, a white-faced petrel from Down Under, thousands of miles from its proper haunts. In all, thirteen species of birds appeared, most of them quite capable of colonizing emergent land.

No amphibians arrived in the vicinity of Beebe's island (nor have they done so anywhere in the Galapagos, over countless years) but almost immediately there were reptiles, two great sea turtles that floated calmly nearby until startled by the ship's waves. Soon insects appeared, first a number of flies and then a score of Cocos dragonflies. Other insects were found dead in hauls of surface nets, and at least one strong-winged butterfly came and inspected the ship and then flew off. In the surface nets also were found seeds of terrestrial plants, and during the ten days at his watery island Beebe saw floating close by two of the great buoyant seeds of the coconut palm, a plant that by one means of dissemination or another has been established on tropical islands around the world. Three other land plants also came into view, two of them rooted to a drifting log, and the third, a form of grass, floating in the manner of sargassum weed. In the shelter of other logs were two dozen shore crabs of four species, and forty shore fishes of eight species—a cogent answer to the question Beebe had pondered earlier about the passage of littoral or shallow-water creatures over wide stretches of deep sea.

Thus had one remote, imaginary island been hypothetically stocked with a variety of living things in a mere ten days, or, on the evolutionary time scale, in a twinkling. Beebe was quite aware of the flaws in his fanciful scheme: the time required to form soil for successful plant life, the likelihood that many of the first colonists would fail to breed successfully, or, having bred, would be eliminated by unfavorable conditions or by predators. Yet he admitted his wonder at the casual largesse of the sky and sea, "this radiation of living creatures, birds and insects, and . . . plants and fish, over half a hundred miles," and confessed that all of this "was rather destructive to former theories I have held." Stated bluntly, he had been proven wrong on important points of his land bridge theory, not by the scientific observations of others but by his own. Many creatures whose appearance far at sea seemed unlikely or even impossible to Beebe in 1923 had indeed appeared in 1925; the oceanic theory had been confirmed by the same man who had previously rejected it. And this man, far from abashed at changing his mind, instead found this new point of view "thrilling" to contemplate.

It is to be regretted that William Beebe's new data and his

abandonment of "former theories" did not receive proper attention from those who wrote later about the Galapagos. Granted, his imaginary island was closer to Cocos than to Tower Island or Indetatigable, but the *Arcturus* expedition and the book describing it were concerned more with the archipelago than with any other location. Even closer to the point, Beebe's "entirely new idea as to the effectiveness of oceanic distribution" was offered in contrast to the land bridge theory put forth in his previous Galapagos work. Even David Lack's elegently stated and persuasive book on Galapagos birds draws only on Beebe's first formulation—which Lack rejects, in keeping with much recent opinion—while neglecting his later implicit disavowal and the data upon which it was based.[12] Ultimately, perhaps, favorable citation by fellow scientists is less important than the recognition that William Beebe advanced beyond his initial conclusions to others of an opposite import, and did so by employing his favorite scientific approach, the study of many living creatures within a relatively small space.

The *Arcturus* expedition, Beebe's second journey to the Galapagos, was also his last; but it was only the first of many explicitly oceanographic voyages to be taken in years to come. By the time the yacht reached its New York berth at the end of July, Beebe had many months of diversified marine experiences behind him, and was well embarked on his new scientific endeavor. To be sure, most of his oceanographic efforts so far had been of the conventional kind—but it was not in the Beebe spirit to remain within the common boundaries, if methods more original and alluring caught his eye and beckoned.

:7 Beneath the Sea

> April 9. [1925] Tried diving helmet for the first time, and found it most exciting experience. Trite but true to say it opens a new world. . . . A large shark was swimming nearby, but paid no attention to the diver. The strange beauty of the submerged scenery is hard to describe.
>
> The Log of the *Arcturus*

With a few steps down a ladder and into the waters of Darwin Bay in the Galapagos Islands, William Beebe ventured into a new realm of nature and entered upon a new phase of his career. [1] On this April morning he stood and gazed about in sun-bright water fifteen feet deep; on an August morning nine years later he would look out into waters two hundred times deeper, and black beyond imagining.

But if this venture was new, it was not unheralded. Aboard the steam yacht *Arcturus* which had brought them to the Galapagos, Beebe and his staff had captured marine specimens in nets on the surface or far down in the Sargasso Sea, and again from time to time at other points along the way, especially in a remarkable rip current encountered three days west from Panama. Two years before, among the tide pools of these same islands, Beebe had collected species both known and nondescript, and had netted and seined and trawled offshore, and had fished with handlines and sports gear—accumulating in the process very many aquatic specimens for study. Earlier by the rivers and estuaries of Guiana he had observed and described various water-dwelling creatures, showing a special intimacy with those inhabiting the zone at the edge of tide. Earlier still on his first great voyage for pheasants into the Orient, he had observed and occasionally collected sea creatures; he relates that he caught a flyingfish that had landed on the deck of the *Lusitania,* and in *Pheasant Jungles* he tells of standing astride the bow of a sampan and "scooping weird things from the swift tide" off the port that serves Kuala Lumpur, Malaya. "Secrets of the Ocean" is the title for one of the most detailed and engrossing chapters of his third book, *The Log of the Sun* (1906). Even as a boy he can be found watching pickerel in a clear New Jersey pond, or seeking out the pools of melted snow where he knew he could find fairy shrimps.

William Beebe was accustomed to writing "Ornithologist" on his passport applications, but in truth the more proper term is an earlier one, Naturalist. Indeed, in America one must look back not to Audubon or Wilson to find a precursor to Beebe, but to Mark Catesby, the English naturalist who spent more than ten years in the southern colonies early in the eighteenth century, and then produced *The Natural History of Carolina, Florida, and the Bahama Islands,* published in London between 1731 and 1748. This sumptuous work, as the title suggests, describes and illustrates many classes of living creatures, foremost among them birds and fishes.

Had William Beebe shifted his emphasis from birds to, say, holothurians, the sea cucumbers, it would have seemed out of character. Not so his shift toward fish—for here in the waters of the tropics were the birds and even butterflies of the sea: demoiselles small and fearless as chickadees, butterflyfish rivaling in sprightliness and color their aerial namesakes, trunkfish as portly and solemn as owls, groupers as importunate as gray jays, surgeonfish as gregarious as starlings, even sharks as forbidding (and generally as harmless) as vultures. And here was color to rival the most resplendent of pheasants or hummingbirds, and grace and swiftness to match the swallow's or the peregrine's. Only the odors and the voices were lacking—although various clicks and pipings now and then come to the ear of the diver, along with the odd grating noise of parrotfish chewing away at coral. And, as with any good bird walk, here too is an ambience within which one moves: not breeze but surge and flow, not bushes and trees but great corals and sea fans and plumes, not flowers but hydroid blossoms and nudibranchs and sea anemones; not thistledown or dandelion parachutes sailing in mid-air, but jellyfish drifting or pulsating in midwater; and, for a touch of peril, not a copperhead or a fer-de-lance to avoid stepping upon, but a poisonous scorpionfish or a sea urchin with painfully sharp and brittle spines.

As to danger, Beebe notes: "It is idle to say that I, and I think all of us who went down, did not feel at first exceedingly nervous." But to love nature is to assume her general benignity, and soon he felt wholly at ease. He was to discover what many a diver and frogman and even snorkeler was to learn also, decades hence—that the principal dangers of undersea ventures are those of forces, not creatures. Surge and current and depth pressure cannot be ignored, but sharks can; the failure of one's own equipment must be taken into account, along with the possible failure of judgment about one's physical capabilities—but a five-foot barracuda is just another big fish, no more to be feared than the giant herring called the tarpon.

What Beebe was here recognizing was not so much the benignity of the salt sea environment as its wonderful indifference to our species. As he notes in *Beneath Tropic Seas,* we spring from "far distant aquatic ancestors," and in *Half Mile Down* he suggests that the salt content of human blood provides a rough gauge whereby we can calculate backward in time to discover "the anniversary of our marine emancipation," when the sea's proportion of salt was like our blood's today. He further states that "fresh water injected into the veins is a potent poison, while sea-water is an admirable temporary substitute for our very life's blood."

Whether these theories are tenable or not, they say something important about human ventures upon and below the sea. The "emancipation" was eons ago, whereas our partial and tentative return to the salt water came, in geological time, only the day before yesterday. In terms of marine ecology, therefore, we featherless bipeds are almost pure exotics. Not so those feathered bipeds called penguins; for at least fifty million years they have made the sea their home, feeding beneath the waters on fish and squid and crustaceans, and being fed upon in turn by leopard seals and killer whales—two other warm-blooded creatures whose progenitors more recently descended from the land to take up life in the sea. Young seals, in turn, must take care not to wander too far out into shark waters; and stricken whales, as Melville and Cousteau tell us, are familiar meat at the shark's undersea banquet.

What then of human flesh at a similar gory feast? In the course of many hundreds of dives, William Beebe never lost his "notorious scorn of sharks," and his death at eighty-four, abed and unbitten, would seem to validate his view. Yet there appear to be many authentic records of men as stray victims of the shark clan's wayward appetites—but Beebe himself once caught a large blue shark by baiting it with boiled potatoes, and there are also records of a sharkish taste for objects even more outlandish. So perhaps here is the answer: *Homo sapiens* as shark fare is an accident, however melancholy, and not an ecological imperative.* Similarly, we plantigrade mammals, unprotected in this new environment by fur or scales or even blubber, may blunder into creatures whose venom sorely injures us.[2] Presumably, however, no organism of the salt sea has yet learned to invade our bodies and kill us, as fresh water organisms assuredly do. For example, certain mollusks of Indo-Pacific seas, especially *Conus geographus* and *Conus textile,* carry potent stings as part of their natural equipment, and in the course of a year a man or two may die by chance encounter with these marine snails; but certain fresh water snails, generically *Bulinus* and *Biophalaria* and *Oncomelania,* carry the larval form of a human disease called schistosomiasis, and in a year's time may bring about many thousands of deaths,[3] as part of the natural fate and environment of man, the earth and the air and the rain water that runs in streams or stands in lakes and pools and ditches.**

As an ecological stranger in a strange and lovely waterland, William Beebe set out as one of the first of scientists to explore it face to face, so to speak. He was well aware that he was not indeed the first to make the attempt; Aristotle himself had made notable observations of marine life by peering into clear Aegean waters, and tradition says that Alexander the Great satisfied his intellectual curiosity by having himself lowered into the sea in a glass-windowed chamber, simply to look at the creatures living there (one of them so enormous that it was three days swimming past the patiently watching monarch). Many other men in many odd devices had also tried, not without success—although too often their motives were acquisitive or martial, not scientific. Beebe was simply the first famous and highly trained natural scientist to take to hel-

* However, given the vastly increased numbers of our species lately taken to the sea—and leaving aside the question of our palatability—we may wonder how long it will take the various mighty predators beneath the tide to learn.

** But again the question of adaptation obtrudes. The innocuous remora or shark sucker has already appeared on the bodies of skin divers; and certain marine parasites have doubtless long afflicted our distant mammalian cousins, the seals and whales—and may one day quietly follow the quick remora's lead.

met diving as a new form of field work. And nine years after his first dive in Darwin Bay he could still report, "The study of life under sea holds, at present, the heart of my mental interest."[4]

Here at the Galapagos his equipment was of the simplest. In New York he had purchased a few sets of diving gear, each including a tall copper helmet, fronted with two vertical panes of glass set at oblique angles from the center. The helmet weighed twenty pounds; for diving, four ten-pound lead weights were added. A hose connected at ear level on the right, with air sent down from a double-action hand pump aboard the attendant boat. In a swim suit and sneakers, Beebe climbed overboard and at about neck level allowed the weighted helmet to be lowered on his shoulders. He stepped further down the ladder, reached the bottom, and began his observations.

In the first moments he became aware of the easeful and impalpable embrace of tropic waters—a temperature so kind as to beguile all sense of wetness, a step so light and flowing as to banish stress; a surge that gently impels fish and fronds and diver alike. Beebe soon found that he should let himself sway with the water's easy motion, for thereby he became only another natural object—whereas if he held himself still by conscious effort he inspired mistrust and flight. And he found that by moving easily along he "became a Pied Piper of sorts, leading a host of fish which followed in my train," and when he tried baiting in the form of a freshly killed crab, he was surrounded by "fish and fish and fish," in such variety and numbers that he felt as Adam did, trying to name all the creatures thus brought before him.

As is commonly true with new fields of nature study, identification was a primary task. Many of these species, whether of fish or various other classes of marine life, Beebe had already encountered in tide pools, or had taken in nets or trawls or by hook and line; others he might have seen in aquariums, or studied in the authoritative works. But many were new, some perhaps only to Beebe, others to science itself. So within the depth range of fifteen to perhaps fifty feet he set about observing and collecting, discovering not only new creatures, but novel methods of procuring them. One such was gathering a tubful of rocks from several fathoms down and allowing them to stand in the sun for a day. Then he found that "an amazing array of interesting beings" had crept from the rocks, creatures otherwise not readily obtainable. Other means of capture included dynamite, admittedly crude and indiscriminate, and spearing with a light trident, which Beebe devised when a heavier harpoon failed to work. With this trident he impaled fish that, to his astonishment, often survived to live in his laboratory aquariums.

How fruitful could be the combination of observing habits through the helmet, collecting by judicious use of the spear, experimenting with specimens transferred to an aquarium, and finally careful dissection, Beebe shows in the chapter of *The Arcturus Adventure* largely devoted to a single species, the surgeonfish *Xesurus laticlavius*. Observations show him the placid herd life of this grazing fish, and suggest that it enjoys nearly complete immunity from predation (despite its good flavor, at least to human taste). Spearing demonstrates again this placidity, since the rest of the school pays no attention as one fish struggles on the barbs of Beebe's weapon. Aboard the *Arcturus*, a controlled experiment demonstrates that the spines the fish carries are indubitably lethal to other species, and dissection shows how fins and organs are placed and structured and utilized to fit the needs of the surgeonfish in its ecological niche. As Beebe notes, however, the relatively technical nature of this chapter is unusual. In general, detailed analyses of these varied inhabitants of the Galapagos shallows may be found in their proper scientific repository, the pages of *Zoologica*. What the common reader gets is the zest of the experience, the "pure sensuous delight" of the undersea world, and he is spared the laborious technical study which necessarily follows, if science is to be truly served.

Although Beebe himself claims "remarkable results" for his scores of helmet dives on this Arcturus expedition, he states that the most valuable scientific returns came not from shallow-water work but from one of the stations set up for study at sea. Designated Number Seventy-four, this station was occupied for ten days and yielded outstanding collections. It was located over water nearly 4,700 feet deep, about halfway between the Galapagos and the nearest point of Central America. In calling it "An Island of Water," Beebe suggested the basic idea of a station at sea: a designated location where the vessel is markedly slowed or stopped, and maintained in position (perhaps by slow circling) over a particular area of sea bottom for the time required to carry out desired observations. Or, in Beebe's metaphor, a station is an island of chosen water in the midst of the greater sea. Usually such a station is occupied for a relatively brief period while a few nets or dredges are lowered, soundings or bottom samples taken, temperatures at various depths recorded. A ten-day pause is unusual; but in the case of Station Seventy-four it was very much worth the time, at least in Beebe's view: "I captured one hundred and thirty-six species of fish in this one spot, and at least fifty species of crustaceans"—the latter group comprising eighty percent of this class collected by the expedition in Pacific waters.

On the first day five silk nets, strung on a single cable at intervals of perhaps a hundred fathoms, plus a Petersen trawl further down, were tried with excellent results. Then came vertical netting, to determine zones of life at various levels. On the third day an otter trawl—a heavy net with boards framing its wide mouth—was sent all the way to the bottom, but came up tangled and with only one specimen of major interest, a grenadier fish. But in the following two days such trawls produced a vast array of creatures, from great black eels to small invertebrates. On the sixth day, however, Beebe lost an otter trawl in trying to put it overside without the crew—"it is their day of rest but the scientist knows none." But this was also the day that he began a twenty-four hour series of surface hauls, putting out successive nets at half-hour intervals around the clock. Not only was the cumulative haul very substantial; the results of the method were especially instructive.

By employing such careful timing, Beebe found that common surface species of the day, such as flyingfish and the gar-like halfbeaks, vanished at 6:30, which in this latitude would be close to dark. At 7:00 six species of lanternfish reached the surface level, followed by other light-bearing species half an hour later. At the end of night the process was reversed, as these species undertook the return migration into the depths, making way for the reappearance of diurnal forms. "After a little practice," Beebe writes, "I knew that if I wanted a certain type of nocturnal surface fish, a haul at 4:15 to 4:30 A.M. would invariably capture some, while a net drawn from 4:45 to 5 o'clock would never contain a single one." Observations of living luminescent fishes, according to Beebe an almost untouched field at the time, were made at Station Seventy-four during these night hauls—admittedly brief pioneer efforts, to be continued in later sea ventures, and reach a culmination in abyssal studies from the bathysphere.

Lest this ten-day period at Station Seventy-four be viewed as an unqualified scientific triumph, however, one should know that the penultimate day was something of a disaster, with one eighty-foot dredge lost and other nets tangled and unproductive, and then heavy squalls and a downpour in the evening. Winds of near-hurricane force howled through the night, so the departure next day from this aquatic corral was not without relief.

Ten days later the *Arcturus* again approached the volcanic eruption along the northwest coast of Albemarle and steamed back and forth while the expedition party watched cascades of lava pour from the shore cliffs into the seething waters below. Then came the failure of the steering gear, fortunately not immediately offshore, but after the yacht had departed from Albemarle and started for Panama on the return voyage. Steering had to be carried out from a wheel aft, but little was required, since the ship made no progress worth mentioning in the next twenty-four hours.

In the six weeks that followed, so many other misfortunes afflicted the expedition that Beebe chose not to include this period in the main body of his book. When the yacht eventually reached Panama, five days were taken up with necessary overhauling—which unfortunately could not include a clearing of her heavily fouled bottom. Once through the canal, the toiling *Arcturus* required six days to cover ten degrees of latitude and five of longitude, from Porto Bello to the passage between Cuba and Haiti. Three days later the shaft of the circulating pump broke and the ship lay dead in the water for many hours while repairs were made. Then, near Bermuda, a falling block laid open the scalp of John Tee-Van, General Assistant to the expedition, and a tetanus shot and crutches were called for when a rusty spike impaled the foot of Jay F. W. Pierson, Assistant in Macroplankton. And again the attempt to find abundant sargassum weed simply failed.

But as Beebe reports in his final chapter, in late July stations were taken up for profitable days of netting and trawling and dredging 100 miles southeast of Manhattan, in the ancient gorge of the Hudson River. From time to time the crew caught sharks, but when the expedition people had shark steak for lunch one day, neither the captain nor any crewman would consent to partake of such a vile dish, though the others found it excellent. Why then, Beebe mused, would they eat raw oysters or fried pork? Another paradox of sorts emerged as they arrived in New York harbor July 30. An illustration shows the *Arcturus,* gleaming white in the morning sun, signal flags aflutter, homeward-bound pennant trailing its 180-foot splendor from the after truck, as she was being welcomed at her Eighty-first Street pier by a crowd joyfully waving handkerchiefs, soft hats, and boaters. But the Log ends on this wry note: "Every ship on the way saluted us, from garbage scows to big liners, and as courtesy required that we answer every blast, we had barely steam enough to creep up to the pier."

A year and a half later Beebe was again afloat in the tropics. This time his vessel was not Henry D. Whiton's 2,400-ton private steam yacht *Arcturus,* made available by the owner's generosity, but the 900-ton sailing schooner *Lieutenant,* chartered by the New York Zoological Society for nearly six thousand dollars. A seventeen-day voyage from New York had brought the schooner to anchor by the Bizoton reefs, a few miles west of Port-au-Prince, Haiti. There Beebe and his staff of nine people took over the vessel as expeditiously as a pirate boarding party, speedily eliminating the crew (they were shipped back home) and suffering the captain alone to remain, to witness the transforming of his ship into a floating biological encampment. A major consideration was cost, and so the basic structures put up aboard ship were of the plainest: tents. At one end of the long deck a big mess tent went up, and at the other end the laboratory. In between were seven tents for sleeping, and tarpaulins were rigged up over deck runways and special pieces of equipment. In this canvas colony the expedition lived and worked from mid-January to May 23, 1927, with no mosquitoes and few major storms, only occasional rains, and copious quantities of good salt air.

Beneath Tropic Seas, the book Beebe put together to relate his Haitian experiences, begins with an apostrophe to helmet diving, which served not only as a conveyance into a "realm of gorgeous life and color," but also as a method used to attain the primary goal of the expedition, finding and identifying and listing the fish of these waters. Strangely enough, no such list had ever been made. It scarcely needs said that within a hundred days the people of the *Lieutenant* had a catalogue of species nearly as long as that recorded for neighboring Puerto Rico in four hundred years. (The next year in *Zoologica* Beebe and Tee-Van published an illustrated and annotated list of 270 species, with a supplement in 1935 bringing the total to 324.)[5]

Beebe's second objective was "to study at close range and at first-hand by means of a diving helmet the life of a coral reef." This was a significant step beyond the relatively brief and scattered studies he had been able to make at Cocos and the Galapagos; here within a few months he made some three hundred dives. He tells us about several of them, bravely essaying a task he knows is foredoomed, the imparting of coral-village reality. When Thoreau contemplated the life of the Concord villagers, he whimsically compared them to rodents sitting at the mouths of their burrows, or scampering to the next burrow to gossip. But here in place of Concord's elms were great elkhorn corals, the streets were sand or ooze or natural rubble, the houses were sponges and furrowed coral heads, and the diverse inhabitants crawled or swam or pulsated or peered out at every level. And to the tall and outlandish visitor moving about beneath his ascending cascade of bubbles, each inhabitant was worthy of study, the commonest events were interesting and the special moments fascinating. Then toward the end of his study Beebe assumed the sober role of census taker, classifying the fishes by occupational status: Free Nomads such as sharks and groupers, resident Villagers such as demoiselles and royal grammas, Sand Crawlers such as flounders, Grazers such as parrotfish and angelfish, and so on—in all, more than forty types inhabiting or wandering through this coralline dwelling place beneath the sea.

No faithful Beebe reader should be unsettled by the discovery that the last two chapters in *Beneath Tropic Seas* are called "Hummingbirds" and "The New Study of Birds." Not till the sun excludes William Beebe will he exclude the birds, and quite without apology he turns to them here. The hummingbird chapter scarcely refers at all to Haiti, although tropical America is invoked as the environmental setting for most of the species. (Beebe claims there are "full five hundred different kinds" in the family *Trochilidae,* but more recent studies limit the species to three hundred and nineteen.)[6] The chapter is a familiar essay underlain by loving incredulity at such astonishing creatures, and it ends *crescendo* with an aerial courtship Beebe witnessed in the Guianese jungle, the two oblivious birds performing their nuptial dance scarcely six feet from his enchanted eyes.

"The New Study of Birds," for all its light tone, is an earnest inquiry into the state of bird study among people "whose lives on earth are brightened by a conscious awareness of bird life," but who do not make a career of ornithology. Beebe recognises that "the lure of the list"—Roger Tory Peterson's apt term—eventually fails to satisfy. Beyond mere identification, Beebe suggests, there are further reaches to attain. One is the intensive study of a given small area, a favorite Beebe procedure. When the species and their relative numbers have been discovered, the reasons for this particular pattern of occurrence may be sought. Or within this population one may set up special rubrics—voice, color range, the diurnal round of waking, feeding, resting, singing, calling, going to roost. Again, individual characteristics within common patterns may be discerned, and close study of form and function can be accomplished with captive birds or wild birds tamed by food and shelter offered within range of powerful binoculars. Banding is another way to check on habits of individual birds; at the other end of the scale, entire flocks may be studied in migration time, perhaps even by watching the face of the moon through field glasses to detect night migrants as they pass overhead. But whatever new approach you decide upon, says Beebe, do not delay too long. Anticipating *Silent Spring* by thirty-four years, he sorrowfully predicts a time not far distant when human devastation will have left bird students with little indeed to study.

Yet one vast region of natural life will remain inviolate, even unknown. As destruction proceeds on land, "the bottom of the sea will be the only place where primeval wilderness will not have been defiled or destroyed by man. He may sail his ships above, he may peer downward, even dare to descend a few feet in a suit of rubber or a submarine boat, or he may scratch a tiny furrow for a few yards with a dredge: but that is all."[7] So Beebe had predicted in 1906—and within twenty years he had taken his first undersea steps and had scratched many a furrow with his dredges. But in spite of his own predictions, that was certainly not all. He was soon to venture far into the ocean's depths, and if he was not to reach the bottom of the abyss, he was to point the way for others to do just that.

On his Haitian expedition Beebe doubtless learned to appreciate the benefits of a fixed base for ocean work, free of such vexations as a dearth of fresh water, a broken steering gear, or a fouled bottom. In 1928 he obtained permission from British authorities to take his tropical research group to the steep little island of Nonsuch in the Bermudas, inaugurating eleven years of study (interspersed with an occasional sea expedition) which would end only with the approach of the Second World War. Here on Nonsuch he established his headquarters and laboratory, well isolated from common distractions but within three miles of the harbor of Saint Georges, where his power launch *Skink* and sea-going tug *Gladisfen* could be docked, along with the heavier vessels soon to carry the famous bathysphere.[8]

And here Beebe could set up a sea station not merely for ten days but for months and years of concentrated study. The spot chosen, a circle about eight miles in diameter over water six to eight thousand feet deep, lay about ten miles south-southeast of Saint Georges. In addition, Beebe discovered a small coral reef area conveniently located near Gurnet Rock, and many other areas for helmet diving in the regions north of the main islands. Although the Bermuda group is situated at about thirty-two degrees north—the latitude of Savannah, Georgia—it benefits from warm southerly currents, and its coastal marine life is rather like that of the Caribbean.

From their shallow-water studies Beebe and John Tee-Van published a substantial field book on Bermuda's shore fishes,[9] and within its eight-mile circle the expedition made some fifteen hundred hauls in the first five years. As Beebe stated, "The job I had set myself on my new island home was the study of the fish of the deep sea and the shore."[10] Not merely the field book just mentioned, but two elaborate monographic studies were expected to emerge from this work, life histories of the fish of both the shallows and the depths. Neither of these studies appeared in the form anticipated; furthermore, a change was in the making which would alter radically both Beebe's method of deep-sea work, and the public notice taken of his enterprise.

As an oceanographer and student of sea life Beebe had gone as far as conventional devices could take him by the time of his Nonsuch efforts. With all the refinements possible in techniques of netting and dredging and sounding, and despite the special concentration Beebe achieved by working month after month in an eight-mile circle, the method itself remained basically unchanged. It had been half a century since H.M.S. *Challenger* had returned from its Homeric oceanographic voyage, having discovered thousands of new species of marine life and hundreds of new genera—by much the same collecting techniques at sea stations that Beebe was still using. And Beebe understood how narrowly restrictive these methods finally were. Imagine, he suggested, the proverbial Martian visitor flying two miles above a fog-shrouded earthly city, lowering some kind of dredge to pick up odds and ends from one or another street; the results would be analogous to those obtained by our dredging of the ocean deeps. What information Beebe had so far obtained scarcely equaled that of a student of African mammals who has managed to capture a few rats and mice but knows nothing of the vast antelope herds, the pachyderms, the great cats.

So if man's devices work so ill, then man must go himself. The question is how—and the first answer had been given thousands of years earlier, by men diving for pearls or sponges, unaided except perhaps for primitive goggles or even an extra supply of air (such as Aristotle mentions) carried below in jars held carefully inverted. Aristotle himself, so far as we know, did not go under water to make his marine observations, and the story of Alexander the Great's fish watching has too much of the fabulous (and too many versions) to persuade belief. A similar problem of credulity clouds the mythic saga of Beowulf, the Nordic hero whose diving apprenticeship occurred when he and another youth swam far to sea on a sort of dare. After sporting in the waves for several days, Beowulf was separated from his rival and then dragged to the bottom by sea monsters, nine of which he slew before returning to the surface and swimming to shore. Another undersea exploit, Beowulf's hour-long swim down to the lair of Grendel's mother, his desperate battle with that fierce creature, and his triumphant swim back, is better known but equally liable to the charge of exaggeration.*

The fascination that the abyss has long held for adventurous men is plain enough, and for many centuries they have tried to fashion devices which could take them beyond the brief moments, the blurred vision, and the relatively shallow depths available to unprotected divers. But whether they planned diving helmets or complete suits or bells or even submarine vessels, the state of their technology held them back. Glass of fair quality and adequate dimensions had been known from early times—but not rubber in usable form. Instead, for air tubes and helmets and mantles the common material was leather; in one version of the Alexander story, the chamber was made of glass and the skins of asses. For diving bells, iron or bronze might be used, but large steel castings are a development of the industrial age. Any metal bell capable of holding one or more people was likely to be quite ponderous, while lighter materials such as wood or fired clay or again, leather, could withstand only moderate pressures. All of these limitations applied to submarine vessels as well, with such things as proper motive power and lighting as added perplexities.

But it is clear that much expert understanding and an abundance of ingenuity informed many of these efforts to carry men,

* Among other legendary figures, the anonymous American hero celebrated by Bessie Smith should not be forgotten: "He's a deep-sea diver with a stroke that can't go wrong,/He can touch the bottom, and his wind holds out so long!"

singly or in groups, safely beneath the sea (if only to make the sea unsafe for their enemies). Upon the insights of the ancients, the odd and speculative attempts of the Renaissance, and the scientific discernment and empirical persistence of the early industrial age, the twentieth century could build its triumphant technical accomplishments. For example, in the eighteenth and nineteenth centuries such Americans as David Bushnell and Robert Fulton and Simon Lake and especially John P. Holland produced increasingly workable submarine designs; but the famous victory of *Unterseeboot 20* did not come until May 7, 1915, when Beebe's erstwhile conveyance, the proud *Lusitania,* was sent with efficiency and dispatch to the bottom off Ireland, accompanied by nearly twelve hundred of her passengers and crew.[11] And not until 1929 did Beebe's newer conveyance take form: a hollow steel ball designed for a different undersea mission than that performed by the *U*-20, and for a better fate, he doubtless hoped, than his old ship had suffered.

On the basis of sketchy recollections of a conversation with President Theodore Roosevelt, Beebe credits him with the idea of a spherical shape for what became the bathysphere. The obvious advantages of a sphere are two: the greatest cubic content for the least surface area, and the greatest resistance to external pressure among all possible designs of similar size and weight. (The disadvantages are equally plain but scarcely crucial: in a rectilinear milieu of horizontals and verticals and diagonals, the installation of equipment in a relatively small spherical chamber presents difficulties, and the fitting of two adult human beings within such restrictive concavities does not make for comfort.) The ball design, in fact, was not new. It appears that a French engineer named Ernest Bazin had devised a spherical undersea chamber for salvage work about 1865, and in 1913 Captain Charles Williamson of Norfolk, Virginia, built a shallow-water observation ball which his son John Ernest used to make a series of undersea movies, culminating in 1915 with *Twenty Thousand Leagues under the Sea.*[12]

Although Roosevelt may indeed have proposed a diving sphere, it was a young New Englander named Otis Barton who independently took up the idea and then persuaded William Beebe to adopt it. Barton, trained as an engineer, had turned to the study of natural history at Columbia University after coming back from a leisurely world tour in the mid-1920s. In particular he had been intrigued by the glimpses of marine life revealed to him by fishermen of the

Sulu Sea in the Philippines. As a boy he had sailed the waters off Martha's Vineyard, and in 1917 he had devised a box-like diving helmet which allowed him to walk around the bottom of Cotuit harbor on the south shore of Cape Cod. Now he was determined to explore far greater depths in a spherical diving chamber of his own design. But in the fall of 1926 the newspapers announced that Dr. William Beebe had undertaken this very project[13]—not, however, with a sphere, but with a metal cylinder which looked to Barton "like a laundry boiler hanging from the mother ship by a steel cable . . . equipped with neither lights nor telephone wires."[14]

Defective though he knew this device to be, Barton was disheartened by the news of a rival undertaking by the famous Beebe, a person who had already inspired him through his books. However, as the months passed with no further news of the laundry boiler scheme, Barton again took heart and at last secured an interview with the man he had discovered to be "as unapproachable as an Indian potentate, and twice as wary."[15] Once encountered face to face, Beebe was readily induced to support the project—perhaps by the impressive blueprints Barton unrolled on the desk, perhaps also by the fact that Barton, through an inheritance from his grandfather, was to assume the entire cost of construction.

This meeting was in December, 1928. Barton promptly took his plans to the Watson-Stillman Hydraulic Machinery Company at Roselle, New Jersey, where technical details were worked out and the first steel casting was made, weighing five tons. When Barton visited the Bermuda installation, he discovered that the equipment available to handle the sphere was not adequate to the task; hence he had the first casting melted down and a lighter one made. The walls of this second sphere were nowhere less than one and one-half inches thick, and the weight was five thousand pounds empty. Equipment and passengers and the door plate added several hundred pounds more. Barton estimated that the sphere weighed about seven-eighths of a ton when submerged.*

"The tank," as Barton called it, was destined to become one of the most photographed and familiar objects in the history of undersea exploration. The basic spherical casting, with an outside diameter of four feet nine inches, had five solid protuberances—

*A considerably more crucial estimate was his calculation that the second casting would withstand the appalling pressures found at ocean depths wholly unknown to living men. At 3,000 feet the sphere would have to bear 1,300 pounds on each square inch, or about 7,000 *tons* on its total surface.

four short legs at the bottom to hold the heavy wooden skids, and a cable attachment flange at the top. Five apertures appeared in the casting, all but one along the equator line: three window ports on one side and a circular door on the other, plus a small opening near the top of the sphere to take the telephone and electric cable. The door, so-called, had an opening fourteen inches in diameter, barely sufficient to admit a grown man. It was closed by a four hundred pound steel plate held by ten large bolts. In the center of the plate was set a heavy brass wing bolt about eight inches across, which could be unscrewed (revealing a four-inch aperture) without removing the entire door. At the end of a dive, sensitive instruments might be passed out through this small hole, prior to the more strenuous work of unbolting the main plate.

The three window ports protruded somewhat from the spherical surface, with the flanking pair set at converging angles. Eight-inch discs of fused quartz, three inches thick but exceedingly transparent, were set into these ports and bolted into place, giving a six-inch opening. Originally it was planned to fit all three windows with quartz panes, but the port opening was filled with a steel plug for the first series of dives, and was plugged again in 1932 after an attempt to install a third window failed.

A single insulated cable carried both electric and telephone lines, entering through a stuffing box tightly packed and screwed down against the chance of a pressure leak at the great depths anticipated. Breathing atmosphere was supplied from oxygen tanks inside, regulated to release two liters per minute. Racks were installed to hold pans of calcium chloride to absorb moisture, and soda lime for removing excess carbon dioxide. A spotlight with a 250-watt lamp (later 1,500-watt) was set up by the starboard window, to be used to illuminate the depths.

To Bermuda and the Nonsuch Island installation came Otis Barton with the sphere in the spring of 1930, bringing with him also 3,500 feet of steel cable and an equivalent length of the special communications and power line. The cable, its strands wrapped and counter-wrapped in a special way to prevent twisting, had a breaking strain of twenty-nine tons, and weighed about two tons submerged. Hence at their furthest distance under water, bathysphere and cable together weighed less than one-tenth of the maximum load specified for the cable.

Barton had chartered as mother ship the big wooden barge *Ready*, to be towed into position by Beebe's tug *Gladisfen*. Two steam winches installed on the barge deck gave power to handle the sphere during dives, the larger one controlling the main cable through a system of pulleys ending at the top of the massive boom, the smaller one controlling the boom itself. As Barton notes, various problems had to be solved on the scene. Methods of lowering instruments and nets had long since been worked out, but to lower into the black depths a great metal ball with human passengers, keep them secure and relatively comfortable, and return them safely to the surface was an endeavor wholly new.

However, as Beebe characteristically suggests, many air-breathing natural creatures had long ago adapted to underwater life—such expert swimmers and divers as the frogs and penguins and cetaceans, or such remarkable arthropods as the familiar whirligig beetle and the larva of the drone fly and especially the water spider, most of whose life is lived in an underwater home of its own devising. The unaided human diver had not nearly matched such accomplishments, having reached perhaps 150 feet in dives of about three minutes duration. Beebe in his helmet had found 60 feet to be a wise limit; the current record for a full diving suit was something over 300 feet, and for a rigid metal suit, 525 feet. Submarines also had reached only a few hundred feet. Over the millennia since our species had first ventured out upon the sea, many men, in aggregate a host, had gone deeper still, but all were dead. The first men ever to descend from the sea's bright upper day into her endless deep night, and watch, and tell of what they saw, and then return, were William Beebe and Otis Barton.

But these men were seekers of knowledge, not adventurers, and so they sent their conveyance on ahead to test the way. It was well they did. When on the third of June, 1930, the bathysphere was lowered empty to a depth of 2,000 feet, it came up in a great tangle of communications line, which took a full day to straighten and rearrange. In the process of winding the steel cable from its original spool to the drum of the winch, an unperceived twisting had occurred, and the twist had been imparted to the bathysphere as it sank deeper and deeper. The sphere had made forty-five complete turns (or perhaps even hundreds, as Barton claims); meanwhile the communications line, clamped to the cable at 200-foot intervals, had been pulled inexorably around the cable with each revolution. On the upward haul it was not possible to rectify the twist. Instead, the line was draped in loop after loop over the bathysphere itself, and in this ignominious condition the whole affair arrived at last on deck.

To discover if any major damage had been done, on June 6 there was another test dive. No significant twisting of the cable or the attached line occurred. All other equipment also appeared in working order, and only about a quart of water had leaked into the sphere by the time it returned from 1,500 feet.

So for the two divers and their technical associates and for the deck officers and crew of the barge and tug, the moment had come. A position had been taken inside the old circle southeast of Nonsuch, with the sea bright and calm. The bathysphere had been cleaned and dried; the final pieces of equipment had been installed. Unable to think of some aphorism for the delectation or instruction of posterity, Beebe crawled silently and painfully over the door bolts and tumbled inside. Sensing the unyielding quality of the steel, he called for a cushion, but none was found. Barton soon followed. The door plate was swung into place and bolted amid a deafening clangor of hammers on wrenches. Then up from the deck swung the bathysphere, and over the bulwarks and out over the Atlantic Ocean. Inside the men turned on the oxygen and conversed in low tones, already thinking of conserving air. At a propitious moment the order was given to lower away. The sphere struck the surface and sank, but the divers inside scarcely felt the impact, and were sure only when froth and green water suddenly surged up over their windows, cutting off the direct rays of the sun. The time was just 1:00.

Only the drifting upward of watery motes told the divers of downward motion—that, and the changes in color: clear green at a few feet, bright bluish green 50 feet deeper, a shadowing and chilling of the green at 100, a fading of green into pale blue before 300. Then as all but bluish frequencies were filtered out by greater and greater depth, the eyes of Beebe and Barton underwent an odd change. Blue was the only color they seemed to see, but even as they neared 700 feet, with the light greatly diminished, both men thought again and again of the word "brilliant" for the blueness and finally the blackish blueness of the water.

But they had other concerns. Suddenly at 300 feet Barton found that the door was leaking. It was only a trickle, perhaps the same one discovered on the test dive. They watched it and it did not increase. Beebe reasoned that further depth would only add pressure and thereby perhaps reduce the leak, but they could not be sure. They decided to go on. Later when an electric switch sputtered and threw sparks in a brief short circuit, they still descended.

The spotlight, occasionally turned on as they sank deeper into dusk, threw a beam that appeared intensely yellow. There were fishes and other organisms to be seen, but the phenomenon of color made them seem almost incidental. Even as it faded, the strange deep blue illuminated the inside of the sphere as well as the surrounding water. And yet Beebe recorded that when he tried to read type printed in a book, he "could not tell the difference between a blank page and a colored plate." If this was something utterly new in Beebe's experience, it was so for the best of reasons: there had never been anything quite like it in the experience of any man in the history of the race.

When the voice coming down the telephone line announced "800 feet," something prompted William Beebe to stop there. The door was still leaking, but no more than it had been; the atmosphere inside was still fresh—indeed, Barton later recalled that it was so rich in oxygen as to induce a kind of euphoria—and no other peril seemed to impend; but there they stopped and for several minutes gazed quietly through the windows. Then they started up, and broke through the shining ceiling of the sea at about two o'clock.

Prudence again decreed a test dive when they put to sea on another favorable day, June 10. Brought up from 2,000 feet, the sphere was dry inside; apparently the right amount of white lead on the door seat had stopped the leak. However, pressure had forced three feet of communications line into the chamber.* With this difficulty corrected, Beebe and Barton were sealed in and lifted overside for another major dive, scheduled to be deeper than the first. But static noises on the telephone line began at the 150-foot level and continued sporadically for the next 100 feet, at which point the instrument went dead.

One of Beebe's first insights into the psychology of such diving was the discovery that the sound of a human voice coming down the slender wire was far more reassuring than such technical certainties as the strength of the cable or the stout rigidity of the sphere; and now, with the telephone silent and his own voice cut off from the people above, he had a fearful sense of isolation, as

* Barton relates that on another manned dive, 14 feet of this cable entered and pressed itself in coils upon him, eliciting from Beebe only a heartless allusion to Laocoön.

though suddenly forsaken. Quickly he and Barton tried to signal through the electric system, as arranged earlier; and then came a sensation of pressure underneath, and they knew they were being pulled quickly upward. As they splashed into the air and were swung immediately up and over the barge gunwale, Beebe glimpsed through his window the tense anxiety on the faces of his associates and crew, and understood that they had truly borne the brunt of these frightening few minutes.

Because broken wire had caused the trouble, 300 feet of the line had to be cut off and discarded. The next day, without a prior test, the two divers again tried for the deep levels denied them before. This time the descent was to be relatively slow, to permit more deliberate observation.

Beebe's account imparts not only the thrill of discovering various creatures as they swam or throbbed or dashed or undulated past the window, but also the sense of verification and fulfillment for the whole endeavor. The years of ocean work and laboratory analysis were here given a new dimension. Recognition of many species was instantaneous, even though conditions of light (and hence of color) were much altered. Creatures which had always come to the surface dead were here seen in what Darwin calls "a state of nature," and so on this dive as on later ones, data could be assembled that would establish unsuspected facts about abundance and distribution and aspects of behavior. It is not merely fanciful to suggest that Beebe was here becoming his own Martian visitor, no longer drifting above an unseen city and dangling a net, but descending among the inhabitants themselves to look, however fleetingly, into their lives.

At the bottom of this dive, 1,426 feet, again there was a silent pause; and there came over Beebe a sense of strangeness and privilege and felicity, that he should in truth be here in the deep sea as none had ever been before, able to see and know it by its own separate sparks, or by the great beam he sent forth into that blue-black endless dark which had not known such light since the distant ages of its birth.

This was the deepest of Beebe's dives in 1930. Of the total of fifteen made that year, the last four were distinctive in being shallow contour dives among the reefs close to Bermuda's shores. The idea was to drag the bathysphere slowly along near the bottom, pulled by the offshore drifting of the *Ready* overhead. But guidance was not easy, and despite Beebe's hopeful suggestion to the contrary, nothing much came of this attempt to make a ponderous tethered sphere into an investigator of coral crevices and crannies. In 1932 and again in 1934 other such dives were made, and the potential value of an undersea survey vehicle was fairly well demonstrated; but the practical form of such a machine was many years in the future, after Beebe had largely given up marine studies.

Relatively few dives were carried out in 1932, but in two ways, ichthyologically and dramatically, they produced the biggest catches. Some of the drama was fortuitous and even alarming. The bathysphere, which had been sealed up and stored in a shed on a Bermuda wharf, appeared in good condition after superficial scars and stains had been masked by a new coat of paint, and visibility improved with a new window. But it was not—nor, as it turned out, was the steam tug *Freedom,* which had replaced the old *Ready* as mother ship. Unused to such a load as the aggregate burden of steam winches, cables, bathysphere, and her regular coal and water supplies, the elderly *Freedom* failed her first sea trial with unsettling dispatch. Barely three miles from shore she was already leaking faster than her pumps could bail her out. The pumps, in fact, did not work. But before the water rose too high in the engine room, she lurched about and steamed full-wallow back to her anchorage. Patching efforts next day were expedited when divers saw a friendly gray snapper swim up to the tug's side and disappear into the hull; but according to Beebe, the plugging of the hole immured this generous creature forever.

The next leak was discovered in the bathysphere itself. Boasting a newly installed port window, it was sent down empty to 3,000 feet, and came up nearly full of water. A thin stream spurting across the new pane showed that it had been improperly seated in its frame. Now the problem was to release the water held at high pressure inside. The best way seemed to be to loosen the wing bolt in the center of the door. A few turns produced a singing hiss and fine jets of water which shattered almost to vapor at the instant of escape. All people and moveable pieces of equipment were cleared from the deck facing the door; then Beebe and an assistant returned to the bolt, wisely standing on either side. More turns brought more strange sounds and vaporous spray, and then the explosion. Instantaneously the bolt was torn from their hands and hurled by a solid four-inch jet of water horizontally across the deck to crash into a steel winch thirty feet away. A photograph taken at the moment before the jet dwindled shows a bathysphere sitting like some strange Civil War mortar, discharging not a shell but a fantastic thirty-foot white projectile, and surrounded by a cloud that looks like smoke but is in fact a mist created by the outrush of sea water under great pressure. Given the "almost straight" trajectory of the wing bolt, one discovers its velocity to have been about 120 feet per second. "If I had been in the way," Beebe says, "I would have been decapitated." This is a miscalculation, however. William Beebe was a tall man, and he would have been more nearly cut in two, at about waist level.

As though this were not enough drama, a repeat performance was staged on September 17. The steel plug had been returned to its place in the port window, but tightened only with a hand wrench—and again the bathysphere came up from a 3,000-foot test dive laden with cold sea water. Once more the deck was cleared and the wing bolt carefully turned; once again implosion was followed by sudden explosion and a fierce jet of water. Beebe's faked—or, more charitably, posed—photographs of his anxious loosening of the wing bolt (taken after the explosion had occurred) were hardly needful; the events themselves were genuine and dramatic enough.[16]

The last dramatic action of 1932 was no accident, but had been planned for many weeks. The National Broadcasting Company had arranged with the expedition to broadcast a bathysphere dive over its network on a designated Sunday afternoon in September, with a simultaneous shortwave link to the British Broadcasting Corporation. Troubles both meteorological and technical delayed things for some weeks, but then on September 22 the risk was taken, despite a fairly rough sea. Deck preparations and the start of the dive were broadcast between 1:30 and 2:00 P.M., and the last several hundred feet of descent between 3:00 and 3:30. Thus the voice of William Beebe and that of his assistant on deck, Gloria Hollister, reached an audience greater than any single lecture or article or book would ever reach, and under circumstances more dramatic than any of these could command. On that autumn day a British public coping as best it could with economic emergencies under a Laborite Prime Minister supported by a heavy Conservative majority in Parliament, and an American citizenry about to decide between Herbert Hoover and presidential candidate Franklin Roosevelt (and soon to suffer through the worst winter of the Great Depression) had the odd privilege of hearing an American scientist speaking of unimaginable things in the utter blackness of a third of a mile beneath the sea, 600 miles from the troubled continent.

This was the twentieth deep dive for Beebe and Barton, and notable in two respects apart from the broadcast. Ordinarily, without the pressure of radio schedules, Beebe would have called the sea too rough for a dive; but on this day he consented to give it a try. The two men inside the steel ball paid rather dearly for that decision. Now and again from 500 feet on down, the bathysphere bounced and rolled as the mother ship tossed on the rising sea, and at the deepest level of 2,200 feet there was so much pitching that they decided to ascend without delay. Not only was equipment—including chemicals— being tossed about; they themselves had to brace against the curved sides and bottom to avoid a battering.

But they had indeed reached this record level, and in the process had discovered much that was new. Even at the deepest point for 1930, just beyond 1,400 feet, there had been traces of light above; this time, however, at 1,700 feet Beebe was convinced that all light was gone, leaving only "black, black, black." And in the midst of this blackness he started seeing creature luminescence so abundant and so active that he began to feel a sense of bewilderment. Amid all the light and movement he resolved to try to discern individual creatures, to connect one point of light with another—in short, to "see" where there was no light except that emitted by living organisms. The spotlight of course could be used, but some species fled the beam, and the luminosity of others paled in its rays. Lanternfish and hatchetfish and round-mouths and the saber-toothed viperfish, all forms which Beebe knew from earlier studies, were identifiable by their photophore patterns. A school of shrimps suddenly attacked by two large fish were visible in silhouettes provided by their own emissions. Once an unknown fish poised near the window, its outline revealed by some kind of inner illumination new to Beebe's experience.

Luminescence also radically amended earlier notions of abundance and distribution, obtained by netting in these same waters. For example, light-bearing creatures were unexpectedly common from 1,700 feet to about 2,050 feet, then rare for the next 100 feet, only to increase until at 2,200 feet they were again bewildering in their numbers. The sheer prodigality of the display and the abounding life it revealed gave Beebe a direct proof of the theory he had long held: the slow dragging of nets is a poor way indeed to discover what actually lives in the deep seas.

But on this day in 1932, hanging by a steel string from a tossing tugboat 2,200 feet overhead did not seem the ideal situation either; and besides, the oxygen was not inexhaustible, and the broadcast had ended. This last was perhaps regrettable, because the biggest ichthyological discovery was yet to come. The ascent had barely begun when Beebe saw two fish slowly pass the window, each about six feet long, shaped rather like a barracuda and, like that fish, generously toothed and large of eye. The teeth were luminous, a string of about twenty bluish lights ran along the side, and lights hung from barbels beneath the fish. From these and other characteristics Beebe placed this new species among the Melanostomiatids or sea-dragons, with the designation *Bathysphaera intacta,* the Untouchable Bathysphere Fish.[17]

When Barton and Beebe crawled from the bathysphere into the freedom and sunshine of the deck, they had been inside for almost three hours, and during that time had been buffeted a good deal, although without real injury. But an unwilling companion of theirs, a young spiny lobster, had withstood the entire journey too, not inside but wrapped and tied outside the sphere, in the expectation that depth pressure would crush it fairly promptly, sending forth its juices to entice fish into range of the observer at the window. Not at all; though the pressure on this doughty crustacean rose to about eight tons and then fell back to the comparatively negligible quantity on deck, the creature not only survived but lived on heedlessly in one of Beebe's aquariums.

Named, so far as we know, without ironical intent, the Century of Progress Exposition opened its doors in the worst year of economic distress in the nation's history, 1933. But in its first six months this plucky enterprise along the Chicago lake front sold twenty-two million tickets of admission, and showed a profit. One of the pavilions typical of its sanguine spirit was the Hall of Science, displaying various inventions and contraptions to celebrate the progress of humanity in the so-called conquest of nature. Among the more benign devices were two large spheres, Auguste Piccard's balloon gondola and William Beebe's bathysphere, appropriately the former above the latter, as if to show man's furthest reach upward and his deepest journey downward in the two great fluids of our globe, water and air.

Countless thousands came to stare and wonder and perhaps be taught by these two spheres; but after November 13 there were no more crowds, only custodial people, and then came a crew to start the bathysphere on its way back to Roselle, New Jersey, for overhaul. Gilbert Grosvenor, President of the National Geographic Society and editor of its magazine, had offered sponsorship for another bathysphere venture—demanding, as Beebe crisply observed, "no condition of a new record, which is why I gave it to him."

First, however, the sphere itself had to be thoroughly checked; and robust though it seemed to Beebe's prejudiced gaze, to the cool scrutiny of technological personnel it was well on the road to dissolution. The quartz windows, clear and sturdy enough to the uninstructed eye, were suspect—and they did indeed break under relatively low test pressures. Some of the metal fittings had crystallized and needed to be replaced. Furthermore, the inside equipment appeared so antiquated as to deserve museum display, but certainly no further use by anyone serious about his work. So, with the aid of various donors, "the tank" of former years became, in the words of John Tee-Van, "a more perfectly adapted machine for its unique task."

Beebe hoped that physical observations from the bathysphere might be added to zoological ones, especially in spectographic study and the detection of cosmic ray activity at deep levels. Unfortunately no spectograph adequate to the task would pass through the sphere's narrow door; and regarding cosmic rays, the Nobel laureate Robert A. Millikan regretfully told Beebe that discharge in the depths is too infrequent to give useful results during the relatively short periods of Beebe's dives.

But the resolve to establish a new depth record remained, and it was to be twice fulfilled. As the summer of 1934 approached, the expedition party (including Otis Barton, invited in from Panama) assembled at the Biological Station and the New Nonsuch Laboratories near Saint Georges. On July 5 the renovated bathysphere arrived by sea and was unloaded upon the deck of the old barge *Ready*, where diving preparations began. A month later shallow-water tests were made; then on August 7 the sphere was sent down empty to 3,020 feet. All went well. Good weather came four days later, so in mid-morning Beebe and Barton climbed aboard and once again splashed in and started down, reaching a record-breaking 2,510 feet and returning in just under three hours.

As before, William Beebe pondered the strange alterations of light as the sphere descended, noting at 1,900 feet that the very last assignable color was a dead gray, not violet, as might be expected from the normal progression through the spectrum. The fact that there was any light at all at this extreme depth he attributed to a very bright day and a calm sea. Beebe also found that the electric beam, which had ranged from pale to intense yellow above 1,000 feet, began to shade into luminous gray and then into turquoise at about 2,500 feet, with a border of "rich, velvety, dark blue." At such depths the range of light was about 45 feet, with the beam showing a turquoise cap at its end.

Three notable fish discoveries marked this dive—the two-foot Pallid Sailfin, a hitherto unknown species seen at 1,500 feet and again 1,000 feet lower, the Five-lined Constellation Fish at 1,900 feet, and the Three-starred Anglerfish at 2,470 feet. All three new species were seen long enough and clearly enough to permit their being formally classified, tentatively at least.[18] Many other creatures of uncertain identity came into view (including, as in earlier dives, large shadowy forms at the outer range of the spotlight) but these three, along with the Untouchable Bathysphere Fish, remain the prize exhibits, so to speak. The sailfin was especially distinctive in its size, its odd dead-flesh color, and its very large vertical fins. The constellation fish was astonishing in its rows of yellow lights, each ringed by smaller lights of bright purple.

This record dive of August 11 was followed just four days later with the deepest venture of them all. For the sake of comparison Beebe sought to descend at the precise location of the earlier dive, as nearly as that could be reckoned. In addition, times of entering the water and leaving it were almost exactly the same for the two dives. But despite this attempt at duplication, the experiences were only roughly similar: "The same spot," as Beebe says, "but far from the same visible life." At many levels the press of living creatures was even greater than before—and immensely greater than netting operations in these regions had suggested only a day or two earlier. As the bathysphere approached the former depth record, Beebe saw a creature he facetiously called a MARINE MONSTER—a brownish monochrome shape at least twenty feet long and rather stoutly proportioned, but too near the end of the light beam to be clearly seen. He thought it might have been one of the smaller whales. Then as they descended to the ultimate depth of 3,028 feet, they passed through regions which seemed to Beebe increasingly rich in large forms. Within the last 500 feet he saw at least a dozen fish a yard or more long, and noted at the same time an increase in general luminescence—facts which suggest direct relationships between size and numbers and depth.

But this was a suggestion that William Beebe was not destined either to verify or to refute, because he never reached these ocean depths again. This dive, the thirty-second major descent he had made in the bathysphere, was almost the last of his life. Within a short time the year's schedule of dives ended, never to be repeated in a comparable way. Substantially, the undersea phase of Beebe's career, which had begun at Darwin Bay in April, 1925, ended here in Bermuda with the last dives of 1934. Otis Barton would continue his deep sea ventures, not in the bathysphere (which in fact he had donated to the New York Zoological Society in 1930) but in a new sphere which he called the benthoscope. For another few years William Beebe was to continue oceanographic work on his Pacific travels aboard the yacht *Zaca,* now and then diving in his helmet with all the old enthusiasm, despite his years. But his explorations of the black abyss were over. The Bermuda installation itself was closed with the coming of the Second World War.

Back in 1926 Beebe had written of a form of snipe eel: "These eels were always quite dead when I found them in [nets] . . . and how they live and move and satisfy their appetites in the icy blackness beneath our keel I shall perhaps never know."[19] Yet four years later from the bathysphere he had seen such creatures living and moving, and could describe one of the species well enough that his staff artist could paint it to illustrate both an article and a subsequent book.[20] Similarly, in the nets the more complex types of siphonophores were reduced to broken and tangled tissue, while from the sphere's window they could be seen in all their strange beauty, one species looking "like an inverted spray of lilies-of-the-valley, alive and in constant motion." A few of these hydrozoans he found to be uniformly luminous, others carried individual light sources, and there were those that were visible to Beebe only when they happened to pass through the light beam. To observe such species and many others, the marine scientist no longer had to depend on dead or dying specimens, or at best on short-lived captives of the aquarium; the way had been opened for field work half a mile down.

What was true of individual forms was also true of species in the aggregate. As noted earlier, Beebe was concerned with both vertical distribution and relative abundance of marine organisms, and was aware even before his deep sea explorations how fortuitous, partial, and inconclusive were the data collected by net and trawl, and how limited were other data gathered at the lesser depths reached by helmet or diving suit. To discover what creatures lived in waters below, say, fifty fathoms, and at what level or levels, and in what relative numbers, Beebe had dreamed of something like the bathysphere, and with Otis Barton had developed and used the actual device. And though Beebe knew well enough that he and Barton were merely two men who had observed for a few score hours in a unique contrivance in the deep sea, he envisioned "scores of bathyspheres" which would permit far more comprehensive results than his own pioneer efforts could boast.

Such was the future Beebe hopefully foretold and by his efforts in some ways prefigured. Yet the very thought of fifty or a hundred bathyspheres raises various doubts. Questions of cost and logistics occur immediately: who would assemble such a fleet, who would support the work with men and equipment and an organizational structure? Otis Barton had borne the initial bathysphere costs, affluent donors had contributed components and equipment (for example, the new quartz windows were a gift from Gerard Swope of General Electric, and the new telephone equipment was donated by the Bell system). But other major expenses of a season's diving had been paid by the New York Zoological Society and the National Geographic Society, private groups both, and of limited means. Devising, building, and operating a bathysphere was a private enterprise with basically intellectual objectives and only limited prospects for practical or martial results. Fifty such enterprises were quite out of the question in Depression times, and no such project has developed since.

Yet in recent decades a number of analogous projects have indeed been carried out, both in America and elsewhere. Virtually all of them involved fundamental departures from the bathysphere, while retaining some of its aspects and of course profiting generally from the bathysphere experience. When Beebe conceived the notion of contour diving and met with only indifferent success, his ends were good ones but his means were found wanting. Much the same is true of his helmet diving; the realization that a scientist could observe shallow water marine life intimately and *in situ* was original and rewarding, but the helmet approach had serious technical limitations.

One major defect was shared by helmet and sphere alike: the necessity to remain tethered to a surface vessel, in the first instance for air to breathe, in the second for mechanical support. Independent mid-water suspension was not possible in either case, nor was lateral movement easy and unrestricted. Even depth was sharply limited. At least in Beebe's opinion, ten fathoms marked the outer limit of safety for helmet work, while the weight of the cable itself made too deep a dive hazardous for the bathysphere. Each thousand feet of depth added more than a thousand pounds to the total burden on the cable, and hence to the strain on the deck machinery and rigging, especially in case of a rough sea during the drive.

It was not long before these problems were taken up and solved by others who knew of Beebe's work and wished, as he did, to know more of the undersea world. The same Auguste Piccard whose stratosphere gondola appeared with the bathysphere at the Century of Progress began construction of his first bathyscaphe just before the Second World War, and in 1948 held initial sea trials off West Africa, with mixed results. A test dive to 4,600 feet showed that the observation chamber (like Beebe's, a cast steel ball) and the dirigible-like suspension apparatus functioned well, but the metal envelope containing the flotation material, gasoline, was so fragile that it buckled in moderate surface waves. A few years later bathyscaphes of sturdier construction and more seaworthy design were successfully tested; in the *Trieste* Piccard and his son Jacques reached the bottom of the Mediterranean at 10,395 feet near the Italian island of Ponza in September 1953, and two French navy men in the *F.N.R.S–3* reached 13,287 feet in February 1954 off Dakar.[21] The principle of the free-moving "depth balloon" had been validated and the bathysphere tether cut.

Helmet diving was being superseded at roughly the same time, again mostly by Europeans. The need to move through water unencumbered by an air hose and yet able to breathe had risen over the years from accidents which had flooded mines or tunnels or had trapped submarine crews, and also from adventurous desires to know what might be discovered in water-filled caves or in the sea itself. Building on the work and experiences of others, early in this century the British inventor Robert Henry Davis developed a device known as the Davis Submarine Escape Apparatus. Its basic principle was the rebreathing of oxygen held in a kind of lung attached to the chest, the oxygen itself being purified (somewhat in the fashion of the air in the bathysphere) by passing through caustic soda. Variations of this apparatus were adopted by submarine commands before the First World War, and also used by "frogmen" of the Second World War in carrying out underwater sabotage, intelligence work, and demolition.

Meanwhile, however, Jacques-Yves Cousteau and his associate Emile Gagnan were developing the celebrated Aqualung, involving not recycled oxygen but compressed air carried in tanks on the back and fed through a mouthpiece to the diver wearing a face mask covering eyes and nose. And so was born the postwar free diver, complete to flippers and wrist depth gauge and dagger and, alas, spear gun.* By such means Beebe's helmet world was greatly extended in both range and depth. Operational depths were trebled and even quadrupled—Cousteau considered 200 feet a practical limit for most aqualung work, and states that 300 feet is for emergency situations only—and divers could now venture at will through caves and canyons and coral forests and great stands of sessile weeds. Such divers might even learn to live and work for days beneath the sea, as with Cousteau's Conshelf experiments, begun in 1962, or the American Sealab and Tektite projects of several years later. In short, Beebe's methods of undersea study were ways not taken.

How then did William Beebe view his years of marine work? Zest for discovery, pride in new knowledge acquired or old knowledge revised, satisfaction with the working out of biological problems or technical perplexities—all these appear in his sea books, from *The Arcturus Adventure* of 1926 to *Book of Bays* sixteen years later. Occasionally Beebe discusses the scientific work itself (more often empirically than theoretically) and usually provides in the appendices of his books a systematic review of scientific accomplishments. Moreover, as he frequently states, the end product appears in his reports and those of his expedition associates in the quarterly *Zoologica* or similar technical journals, where they may be judged not by the lay public but by fellow scientists.

Whether he had won himself a secure place in marine science is perhaps a question William Beebe himself could not answer. There is a revealing passage in *Half Mile Down* in which he admits

* This is not to ignore the humble snorkel mask and pipe, which have revealed such wonders to children and the middle aged—and not merely for play, for among others we number in our company a famous ethologist, the august and charming Konrad Lorenz.

that his deep dives could be viewed either as scientifically fruitful or as stunts. Of course he finds for the defendant, himself; yet one cannot shake off the sense that he is not sure of the more disinterested verdict of history. One reason perhaps was the public pother that attended his bathysphere exploits. To his fellow scientists Beebe regretfully admitted that publicity had all come from "indirect reporting" and was "confused in details."[22] While he did not shrink from public notice, Beebe preferred it on his own terms. From the early years of his scientific career he had written not merely for sedate monthly journals but for newspapers and outing magazines; and as he became well known he was likely to appear in various publications not only as the writer but frequently also as the one written about. And if this was the case for his expeditions to the mysterious Galapagos, it was even more true of his exciting ventures into the black depths of the sea. From these he garnered his major portion of that dubious thing called fame. William Beebe still had some thirty years to live, and other lines of scientific work to pursue, but he would always be best known to his common countrymen as the fellow who went down in the ocean in that metal thing with windows and a spotlight, to see what he could find.

A rude justice abides in this decision, as well as a certain sadness. Surely it was part of Beebe's nature to seek both excellence and recognition; and surely he knew, or quickly discovered, both the rewards and the penalties attached to the second of these two. His self-esteem, never in short supply, was gratified; his book sold well; his lectures were popular—and the sponsorship of wealthy persons was more readily forthcoming. When Cousteau applauds "the command of patronage, and the arts of publicity which Beebe was to wield in his historic undersea explorations,"[23] he speaks as a fellow explorer and public figure, well schooled in such arts himself, and altogether aware of their necessity.

But he does not speak primarily as a scientist. There are others who do, however. When the renowned ornithologist Frank M. Chapman brought out his autobiography in 1933, perhaps Beebe wondered what notice Chapman would accord him. In fewer than eight lines he found his answer, though not his name. Speaking of early museum days, Chapman tells of an anonymous "slender lad in knee breeches" who was brought in by a parent (no doubt his mother) for advice in pursuing nature study. Chapman prescribed a "stiff dose" of reading, and states that the lad went on to become "a prominent ornithologist and subsequently, through a gift of vivid presentation, an eminent popularizer of zoölogy."[24] Considerations of professional or personal rivalry aside, this is a slighting comment on the part of the acknowledged dean of American bird science. Essentially it defines Beebe as a scientist with the equivocal term "prominent," but as a popularizer with the emphatic "eminent." And while the anonymity of the passage conceals its intent from the casual reader, it reserves the unkindness of its cut for the initiates of the scientific community.

Very little was held in reserve by Carl L. Hubbs, the University of Michigan scientist who reviewed *Half Mile Down* for the ichthyological and herpetological journal *Copeia*.[25] In the opinion of Hubbs, to presume to describe and assign generic and species names for "animals faintly seen through the bathysphere windows . . . that were thought to be fish" was not merely improper but fraudulent, even contemptible. The earliest of these, *Bathysphaera intacta,* he finds the "most notorious," and cuttingly suggests that the supposed six-foot fish was doubtless two fish swimming close together, each possessing half of the alleged total of lateral lights. *Bathysidus pentagrammus,* the Five-lined Constellation Fish, Hubbs suggests may be "a phosphorescent coelenterate whose lights were beautified by halation in passing through a misty film breathed onto the quartz window by Mr. Beebe's eagerly appressed face. My limitations prevent me from naming that kind of type." Mr. Hubbs completes his scarification by awarding priority of type designation neither to *Half Mile Down* nor to *The Bulletin of the New York Zoological Society* but to the *National Geographic Magazine.* Among certain kinds of professionals a more scornful conclusion is scarcely possible, short of a reference to a Sunday supplement.

The respected John T. Nichols, Curator of Recent Fishes for the American Museum of Natural History, veiled his dubiety toward *Half Mile Down* with equivocal phrasing. Reviewing the book in *Natural History*,[26] he finds Beebe "assuredly less hampered by standard inhibitions than many of his scholarly friends, which frequently leads to his being criticised as well as commended. We think of this when he [scientifically] names his retinal image of a strange fish, when he writes in a manner dramatic rather than meticulous, and most of all when we stand before the iron ball he calls a bathysphere now resting in the American Museum's Hall of Ocean Life." It is then, says Mr. Nichols, that we know on what shelf the Beebe book belongs—a cryptic comment, or perhaps merely a facetious one.

It can be argued that in panting after the telling phrase, Mr.

Hubbs suffers hyperventilation and a concomitant impairment of vision. Thus Beebe's "clearly seen" dorsal and anal fins of *B. intacta* become Hubbs' "(unseen) anal fin," and Beebe's circumstantial account of *B. pentagrammus* is reduced to a "faint vision of a roundish fish (?)" which is perhaps a coelenterate seen through a glass foggily. Or perhaps Mr. Hubbs' reading skills remain unimpaired; he simply doesn't trust Beebe to tell the truth. In his more reserved fashion, Mr. Nichols suggests the same thing; a "retinal image" is to be deplored as a taxonomical base, a prose "dramatic rather than meticulous" is suspect as a vehicle of scientific information. Although Hubbsian spleen may itself be deplored in turn, both men are well within their rights in attacking Beebe at his most vulnerable point, if thereby they can convey their basic professional disquiet. In plain terms, William Beebe to them is something of a fraud, or at best a flamboyant interloper and dabbler in the science of ichthyology. To expose such a man may be a source of personal satisfaction to Mr. Hubbs, or of regretful necessity to Mr. Nichols—but to both it is a sober professional duty. And if neither confronts the more basic questions raised by the bathysphere experience, it is perhaps because neither takes that experience seriously.

Beebe could take comfort in the fact that for every reader of the slighting reviews in *Copeia* or *Natural History,* there were scores or hundreds, perhaps thousands, who read the generally laudatory notices in the more common prints, and who bought *Half Mile Down* and thrilled to its authentic adventures, its genuine accomplishments, its true mysteries. Surely also he must have been pleased to become friend and mentor to Rachel Carson, and then to read in her splendid book *The Sea Around Us:* "My absorption in the mystery and meaning of the sea have been stimulated and the writing of this book aided by the friendship and encouragement of William Beebe."[27]

But regardless of public notice, good or bad, and in spite of professional obloquy, merited or otherwise, one would suppose that the deep joys and challenges of the bathysphere venture would have pervaded and enriched William Beebe's spirit for the rest of his life. Two great earlier experiences, one enthralling, the other somber and ambiguous, had already done so: the expedition for pheasants to the Orient and the journey into war on the Western Front. In books he was still to write, as in prior ones, he reverted often to these experiences—but to the times of the bathysphere, scarcely at all.

Beneath the flags of the United States and the National Geographic Society, Beebe and Barton stand pensively beside the sphere that has just returned them safely from their record dive of 3,028 feet. (NEW YORK ZOOLOGICAL SOCIETY PHOTO)

Back in the world of sunlight and air after a voyage deep into earth's inner space, the sphere is hauled from the surface as the dive ends. (NEW YORK ZOOLOGICAL SOCIETY PHOTO)

In constant communication all the way down and back, Gloria Hollister, shown here during a tranquil moment, asked questions and recorded Beebe's answers and comments. (NEW YORK ZOOLOGICAL SOCIETY PHOTO)

Gloria Hollister monitors the progress of the bathysphere, her face expressing the intensity of concentration often required during these early ventures into the depths. (NEW YORK ZOOLOGICAL SOCIETY PHOTO)

Sealed inside for a dive, Beebe peers through a quartz window that will be required to withstand about 1,300 pounds of pressure on each square inch at 3,000 feet. (NEW YORK ZOOLOGICAL SOCIETY PHOTO)

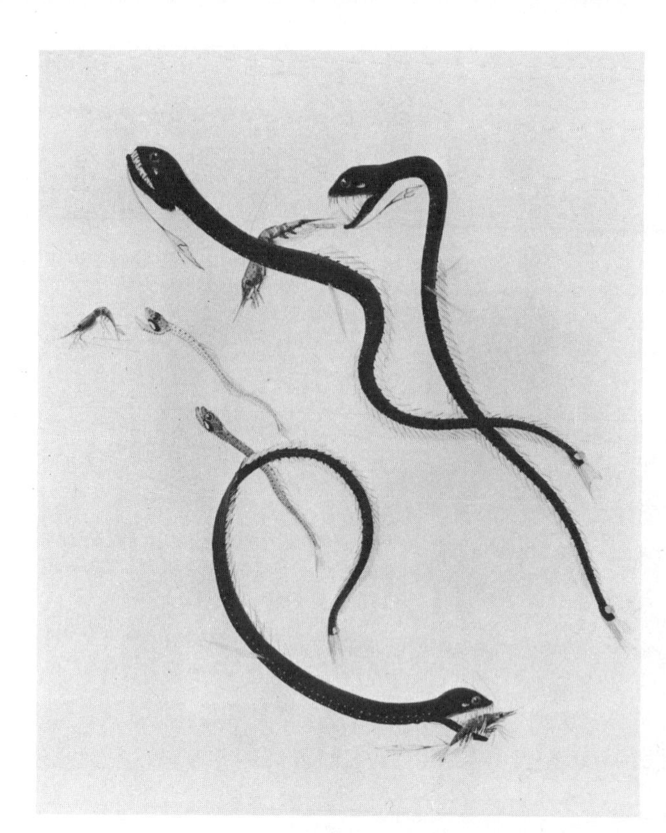

In the depths the Gleaming-tailed Serpent Dragon *(Idiacanthus fasciola)* preys on a school of shrimp. (NEW YORK ZOOLOGICAL SOCIETY PHOTO; PAINTING BY ELSE BOSTELMANN)

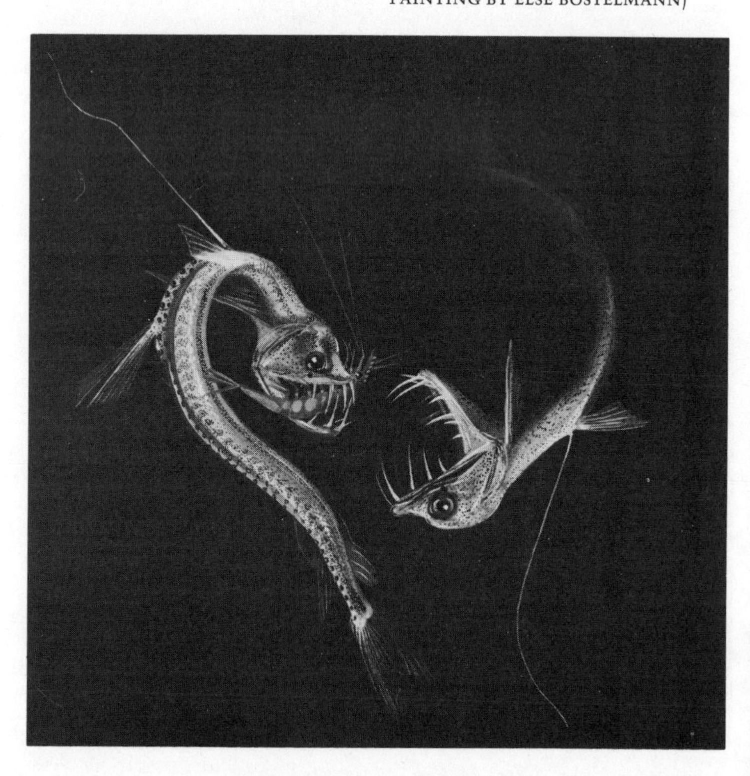

Long feelers trailing, two luminous Sabre-toothed Viperfish *(Chauliodus sloanei)* display their agility in snatching at prey with needle-sharp fangs. (NEW YORK ZOOLOGICAL SOCIETY PHOTO; PAINTING BY ELSE BOSTELMANN)

A deep-sea glutton, the Black Swallower *(Chiasmodon niger)* captures and ingests the Unicorn Fish *(Bregmaceros macclellandi)*, a creature thrice its size. (NEW YORK ZOOLOGICAL SOCIETY PHOTO; PAINTING BY ELSE BOSTELMANN)

The great fish Beebe named *Bathysphaera intacta* swims slowly past the bathysphere beam at 2,000 feet, showing luminous teeth and many blue and golden lights. (NEW YORK ZOOLOGICAL SOCIETY PHOTO; PAINTING BY ELSE BOSTELMANN)

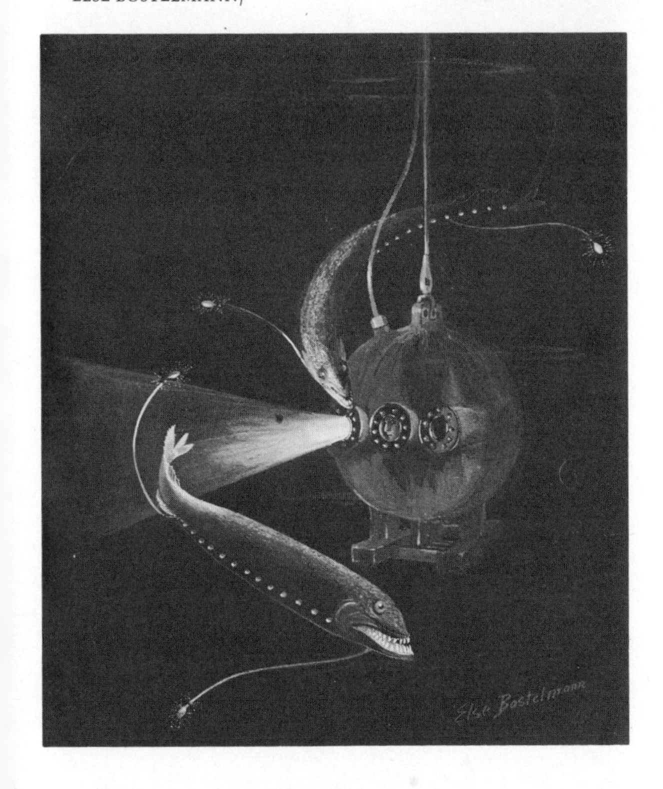

Alexander the Great was an early undersea observer, at least according to the legend, here illustrated in one of its many versions. (NEW YORK ZOOLOGICAL SOCIETY PHOTO)

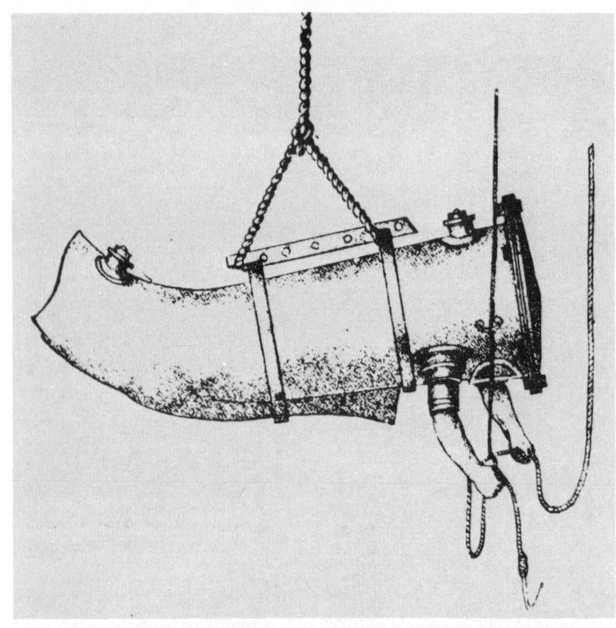

An early salvage vessel, made of ironbound wainscoting in 1715, allowed one John Lethbridge of Devon to search the bottom at depths up to sixty feet. (NEW YORK ZOOLOGICAL SOCIETY PHOTO)

After three hours inside, Beebe crawls stiffly over the rough door bolts and reaches the deck of the mother ship, the steam barge *Ready*. (PHOTOGRAPH BY DAVID KNUDSEN, FROM THE COLLECTION OF DR. WILLIAM BEEBE © NATIONAL GEOGRAPHIC SOCIETY)

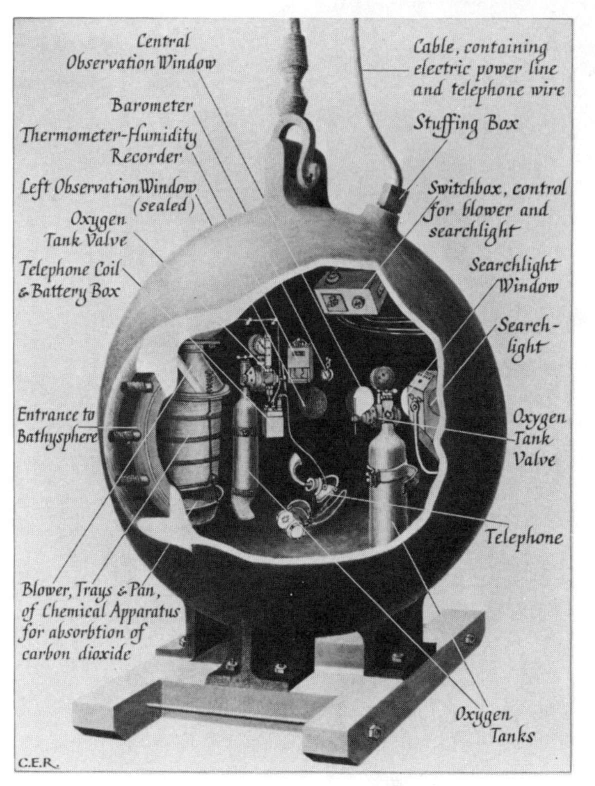

Beebe's famous bathysphere carried two passengers and many pieces of equipment within a steel shell 1½ inches thick and 4½ feet in diameter. (© NATIONAL GEOGRAPHIC SOCIETY)

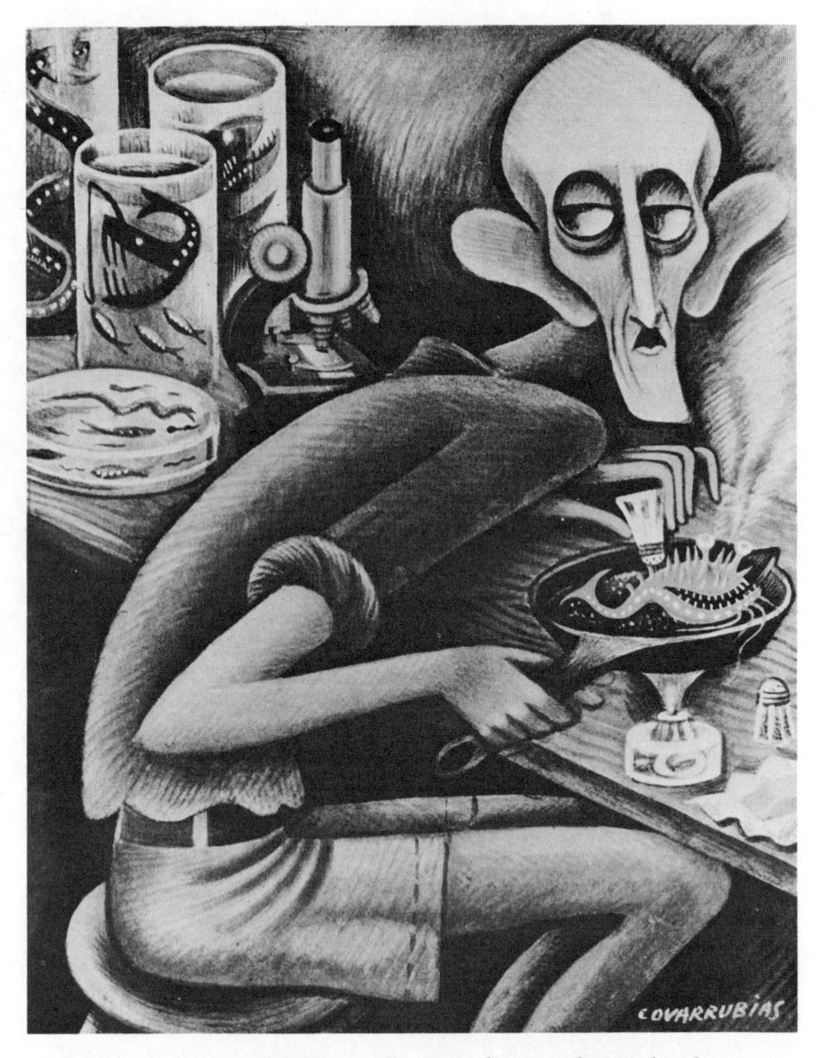

Public notice taken of Beebe's undersea work was widespread and varied. Here the famous Mexican painter Miguel Covarrubias catches Beebe making lunch from a preserved abyssal fish. (NEW YORK ZOOLOGICAL SOCIETY PHOTO)

:8 Females of the Species

> From behind a crag of grey-blue tufa, came one of the girls, swimming lazily, surrounded by all the marvellous fish, with herons and gulls watching her from the nearby lava, and frigates dipping low in flight to see what new fellow islander this was. It was a surprise to realize that she was a mere human, and not what the pool demanded —a mermaid.
>
> *Galapagos: World's End*

The charm of such a small paragraph in the midst of Beebe's writing is enhanced by its rarity.[1] For one reason or another, possibly ranging from vanity or unwitting chauvinism to nice discretion, William Beebe mentioned American women very seldom in his published works, other than professionally. Books quite early or quite late, those involving Mary or his faithful colleague Jocelyn Crane, are exceptions; in between is virtually a masculine desert, relieved infrequently by a fleeting personal reference or a photograph, and even less often by a chapter or an appendix written by "one of the girls." The exorcising of Mary Beebe from *Pheasant Jungles* was a psychic necessity with pervasive literary consequences; the excluding of nearly every other woman from all but professional notice was a calculated policy of Beebe's books for thirty years, with effects less interesting artistically than biographically. Now and then Beebe is explicit about his womanless state: as he sails out of New York harbor toward Guiana he stands perforce alone at the ship's rail; later when he listens to red howlers roaring in the jungle night he has no one this time who shares his excitement. Mostly his life is monkish by implication; women are not noted as being present, or they are but distantly involved.

The reality was perhaps not so cheerless. It is evident from the staff lists of the Department of Tropical Research and from expedition rosters that William Beebe from beginning to end had women associates not merely proficient but faithful—a kind of sisterhood more devoted to Beebe and his work than any but the most loyal of the male staff, in particular John Tee-Van. Whatever combination of attributes a man needs to inspire such devotion, Beebe had in full measure; it is doubtful that any American naturalist before or since has had so many intimate colleagues who were women. Initially they served in his department as artists, and then as assistants in various other technical capacities. Beginning with the two artists at Kalacoon, they accompanied him on nearly all major expeditions and on other scientific trips besides. So the monkishness often implied in Beebe's public writing is something of a gentlemanly delusion.

The two young women who served at Kalacoon did not reappear, but a year later Beebe's party to Guiana included another artist, Miss Isabel Cooper. She subsequently served on expedition staffs at Kartabo and on both the expeditions to the Galapagos, aboard the *Noma in* 1923 and the *Arcturus* in 1925. They are her illustrations which so distinguish *Galapagos: World's End,* in particular the imposing two-page color reproduction of her painting of the land iguana, and the frontispiece, a life-size study of the head of the same creature. On the *Arcturus* expedition she shared artistic duties with Helen Tee-Van, another Kartabo veteran. Mrs. Tee-Van was the most steadfast of all the scientific artists who worked with Beebe, appearing first in *Jungle Days* with an expert and beguiling portrait of the tinamou, continuing with paintings and drawings for *Beneath Tropic Seas* and *Half Mile Down,* and even doing the evocative drawings that illustrated *Reluctant Farmer* (1950), a book by Beebe's second wife. She also became an expert photographer, showing an artist's eye for lighting and composition in many of the seventeen shipboard photos she took for *Half Mile Down*. The single truly indispensable shot of the book is hers: Figure 75, showing the very instant of the explosion of water compressed inside by a leak in the bathysphere.

Miss Ruth Rose appears first on the *Noma* expedition, with the multiple title of Historian, Curator of Catalogues and Live Animals. Her work in *Galapagos: World's End* consists of a long chapter on "Man and the Galapagos," a short one on final collection efforts, and about a dozen pages contributed to Beebe's chapter telling of efforts to discover a source of fresh water in the archipelago. Miss Rose, slender and pretty, shares with Miss Cooper the chance of being the nameless mermaid of the paragraph quoted—and she again shares with Beebe himself the writing of the *Arcturus* book, her title by then being shortened to Historian and Technicist. It had been many years since such joint authorship had occurred; here now was another dark-haired young woman doing chapters for books of Beebe adventures.

The most engaging of her contributions, "Our Islands" in *The Arcturus Adventure,* is something of a lark, a story of exploring two islets with Beebe and "Betty" (no doubt Elizabeth Trotter, Assistant in Fish Problems) between the larger islands of Gardner and Hood. Miss Rose is zestful and far from artless, and she permits herself (or is permitted) to twit Beebe mildly once or twice—and in so doing she suggests a degree of familiarity with her forty-eight year old mentor which he in turn is willing to indulge. The tale itself is told with such high spirits that one nearly forgets the very real hardihood of the two young women and their professional devotion. The chapter "Cocos—A Tale of Treasure" is partly a research piece, as befits the expedition's historian. It is also the story of what the *Arcturus* people found when they came to remote Cocos Island in May, 1925. Miss Rose's section of "Cocos—The Isle of Pirates" is presumably the ten or so pages recounting Cocos tales from as far back as the seventeenth century. She also collaborated with Beebe in the writing of the first and last chapters of the book.

This Galapagos expedition of 1925 was a high point for feminine participation in Beebe's scientific enterprises. Of the five women then listed on the staff, only Mrs. Tee-Van regularly turns up later. In that same year Ruth Rose appeared as coauthor with Beebe of articles in the *Zoological Society Bulletin* and the *New York Times,* but when *The Arcturus Adventure* was published in 1926 her scientific and literary relationship to Beebe came to an end.

Within a year another young woman appeared as Beebe's coauthor with the publication in the *Zoological Society Bulletin* of "The Secret of the Swallow-Winged Puff-Bird" by Gloria Hollister and William Beebe. Miss Hollister's being given primary listing suggests that she was responsible for most of the prose and a good share of the field work (as well as the photographs credited to her) that went into this article, a slight but winning tale of discovering, capturing, and studying the odd species *Chelidoptera tenebrosa.* The scene is Rockstone in British Guiana, about thirty miles up the

Essequibo from the Mazaruni-Cuyuni junction, indicating that with the closing of the Kartabo operation in 1926, this locality was briefly investigated, prior to the major shift of the Tropical Research staff to Bermuda.

On that staff Miss Hollister appeared in 1928 as General Technician, a year later as Technical Associate and then as Research Associate, a title she retained (adding a master's degree in 1931) until she married and resigned from the Zoological Society to take a position with the Red Cross early in the Second World War. During this period she coauthored "The Fishes of Union Island, Grenadines, British West Indies, with the Description of a New Species of Star-Gazer," which came out in *Zoologica* in 1935. This fifteen-page technical contribution resulted from a Caribbean expedition she and Beebe had made in the summer of 1932 aboard the 105-foot diesel yacht *Antares,* owned by Colonel Edwin M. Chance of Philadelphia. (A preliminary report by Beebe and Hollister, describing three new species discovered on this trip, had appeared in *Zoologica* in 1933.) Miss Hollister in 1936 also led her own part of a tropical research expedition to Trinidad and Guiana. Toward the end of her scientific career with the Zoological Society, she became so noted for her methods of preparing tissue for ichthyological studies that she had difficulty finding time for other research.[2]

To the nonscientific public, however, Gloria Hollister was best known for her work during the bathysphere period, particularly as the young woman on the shipboard end of the telephone line connected with Beebe in his steel chamber beneath the sea. Not only were the undersea conversations described and recorded and printed; Miss Hollister herself appeared at her assigned tasks in several photographs in *Half Mile Down* and even more prominently in the *National Geographic Magazine* articles on bathysphere operations. The magazine's photographers, always willing to treat subscribers to pleasant sights, found in Miss Hollister, tall, slim, blonde and notably pretty, a subject calling for candid close-ups, in particular in "A Half Mile Down" for December 1934.

On her birthday, June 11, 1930, Miss Hollister received an unexpected gift—a trip to 410 feet in the bathysphere, accompanied by John Tee-Van. Her account reflects a spirit keenly receptive to the beauties of the experience, and a supple prose style in recording them. Four years later, in the last regular bathysphere season, she descended to 1,208 feet. A photograph she took in those black depths (Figure 91 in *Half Mile Down*) shows perhaps two dozen bright points of light emitted by creatures of the deep sea within range of her lens at the moment of exposure. Here was a highly unusual record of luminescent life, and Gloria Hollister's descent to 1,208 feet was one of the two deepest dives ever taken by a woman scientist.

The second such dive was made by another member of Beebe's staff, Jocelyn Crane, who reached 1,150 feet in the bathysphere that same year, 1934. Miss Crane had first appeared on the Tropical Research staff in 1930 as Laboratory Assistant, and was soon appointed Technical Associate. During bathysphere dives she kept various records of the proceedings, but her more enduring endeavors are shown in the appendix she and Beebe wrote for *Half Mile Down*, "Classified Résumé of Organisms Observed." Another joint contribution arising from this period was "Deep-Sea Fishes of the Bermuda Oceanographic Expeditions," a fifty-page study for *Zoologica* in 1936, and in the same journal three years later Beebe and Crane published a monograph of 175 pages on dragonfish. Such work suggested that Jocelyn Crane was well launched on a career in ichthyology. However, by the time of her second voyage on the yacht *Zaca*, in 1938–1939, she had become engrossed in the study of marine invertebrates, crabs in particular, and she collaborated in writing the next to last chapter of *Book of Bays*, "Dancing Fiddlers." The chapter is gracefully done, and Miss Crane proves herself anything but timorous as she tells of digging with her hands in the black mud of a mangrove swamp for big hard-pinching crabs.

The war years saw the Department of Tropical Research again landbound, with the small staff doing "rainy day" work in New York for most of that period. A major expedition was made to Venezuela in 1942, and not again until 1945. From research at stations at Caripito and Rancho Grande, Jocelyn Crane produced numerous scientific papers devoted both to crustaceans (in particular fiddler crabs) and salticid spiders, the latter study culminating in a three-part *Zoologica* paper on the comparative biology of this group. In addition she and Beebe contributed a paper on the ecology of the subtropical forest of the Rancho Grande area. During this time she was also becoming highly skilled in photography. In Beebe's next to last book, *High Jungle*, she displays her fine talent in almost every illustration—but fortunately not in all; Miss Crane herself, trim and comely, appears here and there also.

This book and two earlier ones, *Zaca Venture*[3] and *Book of Bays*, show William Beebe relaxing the rigor of his monkish pose and speaking less of "Miss Crane" and more often simply of "Jocelyn." Without becoming over-familiar, he lets it be known that she is his faithful companion as well as his colleague. As the scientific world knew Jocelyn Crane increasingly for her researches, the reading public knew her as the closest friend of Beebe's later years and ultimately as the one who attended him in his decline. As C. Brooke Worth says in his thoroughly engaging book *A Naturalist in Trinidad,* in those last years she cared for Beebe "as devotedly as a daughter."

The winning qualities of the man Will Beebe, those attributes often embraced by the term "personality," may serve to explain much of the attraction that women felt toward him. Whether writing or talking of this man, scarcely a person who knew him fails to emphasize his charm, his personal force. Enthusiasm, irrespressibility, a wit both incisive and merry, an open and direct responsiveness to experience at many levels—such is the prevailing testimony. An early story pictures Beebe riding swiftly around the Bronx Zoo on his bicycle, accosting startled visitors with the urgent request that they proceed at once to see his new exhibits at the bird house. Another relates how Beebe and Paul G. Howes, invited on a cruise aboard a government launch in Guiana, discovered a new frog on the canal bank and leaped forthwith into the muddy water to capture it, to the considerable astonishment of colonial officials and their wives. (Faced with a lack of regular alcohol or formalin to preserve such specimens, Beebe simply used gin; but as a man by no means averse to drinking—and so little reconciled to Prohibition that he praised the efforts of American bootleggers—he doubtless felt a twinge at the sacrifice.) Years later Beebe himself told of affronting proper citizens even more thoroughly when he and his friends returned to his Key West hotel from a bird walk, tired and dishevelled but in loquacious high spirits, only to be met by stares of reproof from tourists seated about the lobby. "Why, in heaven's name," he asks, "should we be ashamed of enthusiasm?"[4]

Such vitality and ebullience, clearly evident in Beebe's own work and many times confirmed by colleagues and acquaintances, was a major source of his appeal. And his physical presence, the tall and angular form, the bald head and good slightly aquiline nose and often quizzical eyes, appealed to many as the embodiment of his vivid inner spirit. If Will Beebe was not by common standards handsome, he was nevertheless a man of compelling personal attraction.[5]

But all this scarcely offers sufficient warrant for the enduring feminine devotion given him over periods ranging from a few years to three decades. These colleagues of his were not simple people, but women of complexity and talent, not likely to remain for long bemused by charm alone. The notion of conventional Don Juanism is even less plausible, for in its classic form at least, that particular aberration is based on deep unavowed hostilities and a fear of long relationships. It is clear from his own writings that Beebe nostalgically, romantically, even sentimentally yearned for the enchantment of his early years with Mary, even as he sought to exorcise her persisting presence; but long-term associations with female colleagues bestowed more than the stuff of dreams. These were relationships of mutual regard and reward, offering satisfactions not merely to one of the parties but to both.

At one level, travel with a Beebe expedition promised young women adventurous experience and participation in areas long reserved to men. In a sense Mary Beebe was a pioneer in American natural history, a full partner in the quest for wild creatures in strange lands. Lucy Bakewell Audubon did not accompany John James in his wanderings in the wild hinterlands, nor did Maria Martin, who contributed floral paintings to a few of Audubon's plates. Mary's immediate predecessors were such naturalists as Mabel Osgood Wright, Olive Thorne Miller, Neltje Blanchan Doubleday, and Florence Merriam Bailey, whose books notably contributed to bird study and conservation but whose excursions, placed beside Mary's, were fairly tame and decorous affairs. The young women later associated with Beebe took one long further step, participating as single girls in the famous naturalist's voyages and explorations. And if their appearance in zoological journals signaled changes in the scientific community, their appearance in Beebe's books and photographs suggested even more important changes in American society generally.

One example may serve to illuminate the broader pattern of such change. In retrospect it is touching to read Mary's earnest pleas for a split riding skirt and "a little honest tan" and later for trousers (a graceless variety of knickerbockers, according to her photo) as proper for respectable women in the field—and then to see Ruth Rose in the twenties, short-haired and hatless, casually attired in shirt, swim suit, shoes and stockings, and holding up a big land iguana by the tail; or Gloria Hollister in the thirties, dressed in little more than a sleeveless blouse, shorts, sneakers, and a golden tan. Miss Hollister looks thoroughly contemporary, even (in one photo) to the ragged edge on her shorts; Miss Rose looks a bit old fashioned but scarcely odd; and Mary looks antediluvian. A long history of the liberation of women from restrictive conventions lies behind these three photographs, and in this history William Beebe has a certain place, not necessarily as an advocate but at least as a participant.

How willing a participant is a curious question. Beyond the fact that he needed and sought female companionship, and in the process offered women a place on his scientific staff, his record is more than a little mixed. Here and there in his writings he alludes wryly to feminism, and yet as a naturalist he had a sensitive awareness of nature's pervasive sexuality and in particular of the sex roles of natural creatures. In an essay of 1914 in the *Atlantic*, "The Jelly-Fish and Equal Suffrage,"[6] Beebe devotes most of his space to examples of sex differentiation and sex roles among organisms ranging from coelenterates to birds, but includes respectful references to those "many groups of serious-minded human beings gathered together . . . in order that suffrage and sex equality may be proclaimed aloud to a harassed and somewhat unobservant world." Beebe notes that demands for equality tend to be "chiefly political" at this time, but that the ultimate goal is "a new land of sex-democracy." He ends by denying that specialization of function must "presuppose that one half of the world shall be set to dusting furniture while the other half goes stolidly marching off to war. It is evident that specialization itself is not sufficient; but specialization and thoughtful, respectful cooperation between the sexes—this is the true sex equality." In a lighter vein in *Jungle Days* Beebe ruefully confesses that the male tinamou is placidly agreeable to the idea that he should perform almost all nesting and rearing activities, and that in the clan of insects called thrips the males are "wholly superfluous" much of the time.

A more speculative and extended comment on sex roles appears in "Old-Time People," an essay in which Beebe postulates a special function for the female ancestral ape in fostering those protohuman attributes upon which evolution crucially depended. It was she and not the male who embraced and intuitively cherished the glimmerings of self-awareness that came to both; who fought with the fury of her intuition for a weakling baby, strangely venturesome and aware, the symbolic odd one the males would have killed in the natural course of things; and it was another female in a succeeding generation who accepted the weakling as her mate, and bore him the first ape-man, the new form whose emergence had been made inevitable by female stewardship. And here, at the dawn of mankind, "this first ape man found ape women ready: waiting and understanding."

If it is glib and unjust to say that Beebe is merely giving currency here to the old, condescending, and to some, infuriating idea of "feminine intuition," it is because he is not necessarily using the intuition argument as an indulgent explanation for the absence of allegedly more effectual and desirable masculine traits. But he is indeed saying that the male and female have their roles to play in the evolutionary scheme of things, roles not better or worse, but different. From his many years as a naturalist, and his personal history as a male, has come this concept of sex roles—basically, the male who fortuitously innovates, the female who intuitively understands, safeguards, and waits. At the early dawn of human evolution, such waiting was as indispensable as it was quietly heroic; and in the waiting resided and was prefigured "the future of equality, of splendid unanimity of interest and respect" which would characterize sex roles in civilized society at its best.

It is one thing to theorize on a role, and quite another to play it. In the course of his long career Beebe had a great deal of opportunity to evince "equality" and "splendid unanimity of interest and respect" in actual relationships, and in the case of his most steadfast female colleagues he evidently did so. In other cases, beginning with Mary, his performance is not so clear. A man of such complex enterprise and abundant talent may not always be prodigal of the energies demanded by the practice of equality or the winning of unanimity; it is more natural to use the power of authority, particularly if he in fact commands it by virtue of study and experience. It has been mentioned earlier that Mary's opinions are seldom asked, her contributions seldom noted; or, in terms of the evolutionary pattern Beebe suggests, it is William who originates and decides, Mary who offers quiet understanding. Beebe's male associates have suggested more than once that he was a demanding person to work with, and the fact of gender alone may not have made things notably easier for the women similarly subject to his authority.

Though Beebe is willing to discuss the fact of sex roles, and to indicate his own outlook, he is virtually silent on another aspect of sexuality in nature, sexual performance. References to the mating of creatures are few, hasty, and conventionalized; courtship is frequently discussed but its natural consequence, copulation, scarcely exists. A notable exception, the few lines in *High Jungle* on the coupling of parameciums, is a good deal less than an apostrophe to the sexual urge. Following his books done with Mary and their precariously euphemistic celebrations of the flesh, Beebe's writings had so little to say on the subject of sexual performance that once again his own monkishness is implied—and any hints to the contrary sound merely fatherly or avuncular. But if the truth of the matter was somewhat less bleak, William Beebe continued with some success to observe the outward proprieties of his late Victorian bourgeois upbringing, and surely cannot be called a public braggart in the manner of *la mia lista* or *mille e tre.*

According to a man who knew him, Beebe looked upon marriage as "the most wonderful thing in the world," even though he remained single for many years after he and Mary were divorced. Then early in 1927 a young writer named Elswyth Thane visited the Beebe expedition near Port-au-Prince, Haiti. Half Beebe's age, she had been born Elswyth Ricker in Burlington, Iowa, had written for newspapers and motion picture studios, and had published her first novel, *Riders of the Wind,* in 1926. This book, contrasting domestic frustrations with high romantic adventure, was dedicated to William Beebe, and reflected his influence both in its scenes of major action among the great Hills of India's northwest frontier, and in its hero, nicknamed Dodo, tall and lean and intrepid, and considerably older than the heroine. Before the story ends both have killed natives who threaten their lives, he using a rifle, she a knife.

Four months after the Haitian venture ended, Elswyth Thane became William Beebe's second wife, the marriage taking place September 22, 1927 aboard Harrison William's yacht *Warrior* anchored at Oyster Bay, Long Island, with a number of distinguished guests in attendance. The next year the book *Beneath Tropic Seas* was dedicated to her. That summer Miss Thane (she retained her pen name) began the first of many seasons of research in England, for the most part at the British Museum. *The Tudor Wench* (1932) and *Young Mr. Disraeli* (1936) were two of the best known of her works on English historical personages, and both were produced as plays. Beebe notes that he read *Young Mr. Disraeli* on his way by rail across the United States in the early spring of 1936. He was soon to board the yacht *Zaca* for a nine-week voyage along Baja California, and she was no doubt working on her next book, *Queen's Folly.*

Which is simply to say that both careers continued with no major interruptions. He had many more expeditions to go on, and she had many more books to write. In the case of Miss Thane, when the Second World War precluded further trips to England to gather materials, she began a series of books on American themes, the seven "Williamsburg novels" from *Dawn's Early Light* in 1943 to *Homing* in 1957, and then works on George and Martha Washington, beginning with *Washington's Lady* in 1960. On these books, based on extensive research but written in an appealing popular style, her current reputation appears to rest.

In the mid-1930s the Beebes acquired a residence in Bermuda and for six years, off and on, worked to make it their personal retreat. The coming of the war brought that plan to an end also—the Beebe place was swallowed up in a new airfield. Then in 1942 Elswyth Thane found a rundown farm in Vermont which promised to afford another refuge, and when Beebe returned in the early fall from Caripito he offered both approval of the project and some of the money to help it along. Thus began the story which Elswyth Thane published eight years later in the book *Reluctant Farmer*[7] a tale of reclaiming the old place, restoring its land and sugarbush to productivity and its century-old house and neglected barn and sugar works to efficiency and utility, comfort and charm. A great deal of hard and devoted toil went into the process, and from it came much joy. Miss Thane in her buoyant and yet forthright way gives us both sides of the story.

Reluctant Farmer is among other things Elswyth Thane's portrait of a marriage between a famous naturalist of advancing years (Beebe was seventy-two when the book appeared) and a well-established writer of historical fiction. Although the book concerns them both, in subtle or implicit ways it serves to emphasize their separateness, even their independence one from the other. As he grew older Beebe increasingly disliked cold weather, and so the Vermont retreat was generally a place he visited in the warmer months only. During the period 1942 through 1948, Beebe went on four major expeditions, all to Venezuela; then in 1949 he purchased Simla in Trinidad as his own tropical research station. Here he customarily spent about half of each subsequent year, and here he came in 1962 for the last time, dying late that spring, far indeed from Vermont. As he had recommended in the *Atlantic* in 1914, and as *Reluctant Farmer* testifies, Beebe's second marriage was one of "true sex equality," rewarding to both partners but encumbering to neither.

:9 Home from the Sea

> On April 4th, 1925, I was on the *Arcturus*, on my first deep-sea expedition. I find in my journal of that day: "In the afternoon, a Petersen trawl was put down to 700 fathoms, and the best haul of the trip was secured." Thirteen years later, on April 4th, 1938, on the *Zaca*, a few degrees to the north, I wrote: "Rather heavy swell this morning, but three nets went down to 500 fathoms nicely, and brought up an unusually fine assortment of deep-sea creatures, together with several extraordinary new organisms."
>
> *Book of Bays*

When Beebe wrote this passage for the last chapter of his last book on the sea, he may have known that it would stand as a three-sentence summary, a kind of alpha and omega of his career in marine science. When the *Zaca* made port the next day, that career was essentially over. The last major oceanographic voyage had been taken, and from it would come the final book of sea adventures. The Bermuda station continued to operate, but soon it too would close down, a casualty of war. In 1942 Gloria Hollister resigned and John Tee-Van left the Department of Tropical Research for other duties, eventually becoming General Director of the Zoological Park. Just as the seventeen members of the scientific staff aboard the *Arcturus* heralded the enthusiastic beginnings of Beebe's formal marine work, so the small staff of the war years—Beebe, Crane, and one or two others—signaled its end.

But the last couple of sea books, both concerning voyages aboard Templeton Crocker's diesel yacht *Zaca* along the west coast of Mexico and Central America, were in no way valedictory or low in spirits. *Zaca Venture* (1938) and *Book of Bays* (1942) offer again the characteristic Beebe enthusiasm, many scenes of strangeness, beauty, wonder, or excitement, and an abiding sense of satisfaction on Beebe's part in his accomplishments. And there remains a degree of personal risk and daring undiminished from earlier times. Beebe undertakes extended night vigils at the submarine lamp and on the turtle beach of Clarion Island, he climbs along difficult shores and over parched ridges and headlands, and he goes diving in his helmet, emerging in one instance "racked and bleeding" from the buffeting of powerful shore currents—all of this in his fifty-ninth and sixty-first years.

Especially in the second book a subject latent in all his major work appears boldly: concern for threatened species and habitats, and dismay at human devastation. Beebe offers an account of the evolutionary history of marine mammals, noting how whales have reverted entirely to an aquatic environment and have thus unknowingly doomed themselves to extinction at the hands of man. The northern sea elephants of Guadalupe Island, not so far along the evolutionary path but terribly vulnerable when they return—as they must—from the sea to their breeding and nursing grounds ashore, Beebe studies with both scientific curiosity and sympathetic apprehension for their future. Then on the San Benito islands he is impressed by the alertness of the sea lions, hundreds of which scramble down the beach to safety as he and his party of "terrible invading bipeds" approach. It is not until the expedition has ended and he is back in New York that Beebe hears the shocking news that the San Benito seals had meanwhile been slaughtered en masse for commercial dog food. Knowing that a few must have escaped, he cannot abandon hope that the colony may reappear; but he finds in this melancholy event yet another example of the deathly violence which, for momentary gain, we visit upon our fellow creatures on earth.

Beebe is likewise distressed to discover the radical disruption caused by the United Fruit Company in ripping up the Costa Rica jungle to establish vast plantations of bananas, in order that certain men far away might fatten their bank accounts. He had seen analogous forces at work in Malaya and later at Kalacoon, with rubber the product and profit again the motive. Such exploitation of the wilderness was as old as civilization; but, just as Audubon had witnessed and regretfully recorded the process in his time in America, so Beebe a century later took note of its spread in regions more remote, and similarly registered his dismay.

So much of *Book of Bays* is taken up with activities ashore that the title seems almost a misnomer. Frequently the bays visited offered more interesting studies of land creatures—human inhabitants included—than of marine life. To a lesser extent the same is true of *Zaca Venture.* Not since *Jungle Peace* had Beebe written so sympathetically and humorously of local life, and in both books he did so without reverting to the heroics of *Pheasant Jungles.* Once more he shows himself sensitive to mood and spirit, nuance and paradox, and aware of the complexities behind the picturesque exotic surfaces. The chapter title "Night Fiesta in Manzanillo" suggests touristic effusions, but the pages themselves offer wit and sensibility in what by now can properly be called the Beebe tradition. There is nothing in these books of the crabbed or peevish, nothing to suggest that William Beebe was already in late middle age. *Book of Bays* in particular is a work rich in creative energy and imagination and good spirits, and a bridge of graceful design between tropical seas and the tropical land.

Located about forty miles inland from the Caribbean, the base for the Zoological Society expedition of 1942 was Caripito, a town on the Rio Caripe in northeastern Venezuela. Hills and mountains rim the area to the west, north, and east; in those days jungle of a rather stunted character stretched to the south, giving way to savanna and llano; and eastward toward the Gulf of Paria were the same mangrove swamps and forests which the Beebes had first seen thirty-four years earlier. In these diverse areas Beebe and his party of four—Jocelyn Crane, the artist George Swanson, Henry Fleming, and Mary Vander Pyl—studied the ecology of the region, assembled life histories, and recorded the responses of various creatures to both the springtime dry season and the summer season of heavy rain. While these studies were generally satisfactory, Beebe noted with evident regret that expanding oil operations in the Caripito area were rapidly altering the environment and decimating or even exterminating local forms of life. For this reason among others, the next three major expeditions of the Department of Tropical Research, in 1945, 1946, and 1948, were centered at another location called Rancho Grande, in the coastal mountains of Venezuela eighty miles by road west of Caracas.

Rancho Grande, built by the dictator Juan Vincente Gomez both as a showplace and a stronghold against possible armed revolt, was left unfinished at his death. When the Beebe group first saw it ten years later it was already something of a ruin, a decaying modern castle set against a mountain slope just below Portachuelo Pass, in a zone of life Beebe described as "the ultimate cloud jungle, a maze of lofty flowering trees, giant ferns, portieres of vines, air-plants clinging to trunks and branches"[1]—an exciting area for study in the sub-tropical minor Andres. The regular scientific party of Beebe, Crane, Fleming, and Swanson lived and worked in a finished portion of the vast structure as guests of the Venezuelan government, whose generosity may have been prompted in part by potent American oil interests with links to the New York Zoological Society.

Although the regions roundabout were in many ways quite different from the richly tropical Kalacoon-Kartabo lowlands, the published results, whether found in technical papers or the popular book *High Jungle* (1949), similarly embraced a wide range of life. Jocelyn Crane, as noted, published a three-part study of jumping spiders, a paper on fresh water crabs, and subsequently two more sections of the spider monograph. Henry Fleming offered studies of several insect families which included his discovery of new genera and species of moths; other researchers used Rancho Grande collections for papers on creatures as diverse as ectoparasites and mammals. And in addition to the various living things described in *High Jungle,* Beebe published in *Zoologica* notes and articles on regional ecology, birds, fish, snakes, and insects. Back once again to jungle studies after a long absence, he claims himself to be a generalist still:

> What I feel deeply, by hunch or conviction, is that in my particular case it is from steady concentration on a variety of living organisms and their constant comparison, that, if ever, a glimmering of truth will emerge. Meanwhile, even if this is absolute fallacy, I will be living the most superb existence of any human being in the world (so I think). Many of my fellow scientists have a shorter word for it.[2]

Beebe's awareness of criticism for his casting a wide net in natural science was not new, but his overt response in this book was sharper than before, while its "variety of living organisms" showed him unrepentant. No doubt some of these critical "fellow scientists" were even more put off by many pages of *High Jungle* not devoted to professional matters at all, but to various living arrangements at Rancho Grande and nearby Maracay, where the party stopped to await the installation at the castle headquarters of a thousand pounds of books and scientific equipment. Beebe offered homely details of the rooms he occupied, the window and balconies, the bed, the shower, even the toilet paper perforated by *Dermestidae* to make a fanciful kind of player-piano roll. Such cheery domesticity reminds us that as Beebe approached old age he was increasingly grateful for the amenities offered by civilized life, even while he continued with undiminished zeal his quest for knowledge in the wild.

At least in one instance, involving a sloth, his enthusiasm laid him low. "St. Francis of the Plaster Cast" is Beebe's rueful title for a chapter describing a mishap with a ladder which broke his leg, and the inactivity it imposed. But "forced immobility" is a better term; William Beebe awake and truly inactive is not easy to imagine. With his leg in a cast he was forced to sit still, but he soon saw to it that his canvas chair was transported into the nearby jungle, and himself with it. There he sat for many days of his convalescence, actively engaged with the life about him through binoculars and hand lens and naked eye, and nose and ears. The advantage of such immobility was quickly apparent: he became so much a part of the scenery that wild creatures scarcely noticed his presence. One morning a five-foot boa constrictor soundlessly emerged from the greenery, approached and played its quick tongue against Beebe's leg cast, then moved sinuously by and was gone. Later a hummingbird, perceptible only in the whirr of its wings and the silhouette it made, perched for a moment on Beebe's hat, and, departing ungreeted, left behind a tiny calling card.

Another day Beebe was taken to the beach far below—where, according to the story, Gomez the tyrant had kept a yacht in readiness for escape from insurrectionary countrymen—and again sat in his chair and watched, unable to venture out into the crystal shallows to meet his old friends the parrotfish and demoiselle and wrasse, or even to escape a drenching from a tropical shower. That he should sit still by necessity and not by choice is for him, as for the reader, a novel experience; and yet it is the old Beebe still, critically aware of life about him, observing, speculating, celebrating.

One does not have to read more than partway through *High Jungle* to discover that Beebe's attention is increasingly directed to the smaller, less vivid, less assertive forms of life: amphibians, arthropods, the lower invertebrate orders. Partly this results in close studies not unlike the analysis of a few square feet of Amazon jungle humus Beebe had conducted thirty years earlier; and in particular it reflects a growing preoccupation with insects. The plethora of insect life in the cloud forest had much to do with the choice, plus the fact that the mountain pass just above Rancho Grande provided a migration route (especially for butterflies and day-flying moths) without parallel in Beebe's experience. *High Jungle* demonstrates that the study of ants and beetles and moths and butterflies had taken stage center in Beebe's mind. With special emphasis on the latter order, the Lepidoptera, his *Zoologica* papers from Rancho Grande are similarly dominated by this class of creatures.

High Jungle, the last of Beebe's major books, gives an initial impression of relaxation, even insouciance, suggesting an author serene in spirit and well content with his life work. The form is still the essay, the style is still genial and familiar, but now there is a more colloquial vocabulary and a broader use of contemporary references—kibitzer, hitchhiker, blurb, double take, cushie; Gene Tunney, Edward Everett Horton, Walt Disney, Edgar Bergen and Charlie McCarthy, John Kieran and Franklin P. Adams of "Information, Please." This trend leads here and there to a certain verbal self-indulgence, or a jocularity bordering on apology: phrases like "if the reader is still following this interminable digression" imply a faltering of the magisterial confidence which earlier had been an unspoken central motive of his writing. Perhaps his recurrent defense of his generalist approach reflects the same thing. In his long life Beebe had seen the ranks of the biologists swell not only with specialists but with doctors of philosophy trained in the universities to narrow and more rigorous scientific disciplines than Beebe would ever have been likely to abide. With the fame which had come to him for a half century of devoted work, Beebe was unchallenged as America's foremost naturalist; but he sensed that the day of the naturalist was passing.

And so a more attentive reading of *High Jungle* imparts dismaying hints of personal doubt and protestation, reaching an unfortunate climax near the end with "Jungle Weapons," intrusive in content and style, over-assertive and even truculent in tone. This chapter offers a set of comparisons between the defensive or offensive techniques of natural creatures and those used in wars of man. Such analogies are at best shaky and at worst strained and foolish—and they were just the sort of thing Beebe had reprehended in "Battlefield of the Shore" in *Nonsuch: Land of Water.* "Jungle

Weapons" is generalist with a vengeance, and can be explained more in terms of Beebe's personal crotchets and impulses than his scientific learning.

So jarring and atypical a chapter is fortunately not permitted the last word. The next chapter but one, "Palisades Magic," is among the best short studies William Beebe ever wrote, succinct, assured, directly felt and stated. The final chapter, devoted to North American migrant birds, gives *High Jungle* a subtly muted ending, and with the departure of the birds for distant homes a wistful sense of springs long past.

In 1948, a few months after the Department of Tropical Research expedition left Rancho Grande for New York, Venezuela terminated a dozen years of democratic experimentation by a military coup from which Colonel Marcos Perez Jimenez emerged the dominant figure and future dictator. Though Beebe had no great interest in political matters, the unsettled state of Venezuela was not to his liking. The following year the tropical research group did not take a major trip, but inquiries and visits were made to discover a new base of operations. A fine old house in Trinidad's northern mountains was the ultimate choice, a place given the nostalgic name of Simla, after the town just below the great Hills in northwest India where Kim had been initiated into the mysteries of the Great Game, and where Beebe had come on his own great venture in 1910. The house, a large and elegant bungalow, had originally served the governor of Trinidad as a weekend retreat; later it was owned by the prominent Siegert family, and then had been used as a place of convalescence by the U. S. armed forces during the Second World War. Initially Beebe acquired only the house and twenty-two acres of forest in which it stood; a later purchase brought the estate to more than two hundred acres, and in 1953 Beebe donated the entire holding to the New York Zoological Society for one dollar.

The island of Trinidad, at that time a British colony and presently, with nearby Tobago, an independent nation, offered many advantages besides a relatively stable politics. In contrast to the Lesser Antilles, which curve northward along the rim of the Caribbean from Grenada and Barbados to Puerto Rico, Trinidad and Tobago are in geological terms part of South America. Hence their biological heritage is continental rather than insular, and correspondingly rich—embracing, for example, more than 400 bird forms and no fewer than 617 indigenous butterflies.[3] Although Trinidad has both an industrial and an agricultural economy, much of the island remains forested, especially the steep valleys and crowded minor peaks of the Northern Range. Two miles directly north of Simla stands 2,800-foot Morne Bleu, flanked by montane rain forest and crowned with what Beebe called Elfin Woodland, an almost impenetrable tangle of stunted trees draped with moss and shrouded much of the time in mist and rain. When Beebe came to establish his station the lesser slopes were clothed in lower montane rain forest, undisturbed at least in the sense that such areas were not under regular cultivation.

Yet a good deal of the lush green cover of the Northern Range was something less than the forest primeval. Simla, located half-way up the valley between the town of Arima and the summit of Morne Bleu, stood on a slope overlooking not jungle but cocoa and citrus plantations. The great trees that seasonally burst into orange-pink bloom were of the species *Erythrina micropteryx*, the mountain immortelle, an exotic from Peru brought to Trinidad to shade the cocoa. As Beebe reported in *Zoologica* in 1952, the immortelle had since engendered and sustained a rich ecology of its own, and in addition some of the cocoa and citrus holdings were so haphazardly cultivated that they had reverted to the wild state. As in Guiana, the area adjacent to the Tropical Research Station had been subjected to varying degrees of human exploitation; what was lacking here in Trinidad was the true jungle nearby, stretching away for endless green miles.

As a result, then, of the restricted environmental conditions of the Simla area, Beebe and his party might expect to see the ocelot occasionally but not the jaguar or the puma; the introduced mongoose but not the great anteater, the red howler but not the beesa monkey and only rarely the capuchin, by then nearly extirpated; the fer-de-lance and the bushmaster, but, for lack of swampy habitat, not the anaconda—though that great snake is recorded from the Nariva Swamp on the eastern coast of the island. The shallow Arima river might harbor an occasional small caiman and a few species of fish, but it did not provide the depths and reaches for the alligator or giant catfish or perai or manatee, nor the conditions favorable to the capybara and the tapir. Nature at Simla still offered many a challenge to the scientist, but on a modest scale, in keeping with accessibility and comfort. For his last tropical station Beebe had chosen a place which could be reached by nonstop jet and automobile in about five hours from New York's Idlewild airport.

The announced aim of the Simla operation was to study the adaptive behavior of creatures living in the wild—an enterprise directly related to earlier projects in Guiana and Venezuela—and to work on long-range questions concerning evolution in birds and various arthropods. Aside from an ecological survey, however, Beebe's scientific papers from Simla were devoted to insects only. The last of them, a collaborative effort of forty-three pages called "A Comparison of Eggs, Larvae and Pupae in Fourteen Species of Heliconiine Butterflies from Trinidad, W. I.," appeared in *Zoologica* late in 1960. Two years earlier Beebe had published "The High World of the Rain Forest" in the *National Geographic,* with a relatively brief text but many vivid color plates, again mostly of insects. Although during the Simla period he produced some popular writing in other areas, it is clear that Beebe, who began his scientific career as an ornithologist, ended it absorbed in entomology.

His coauthors for *Zoologica* papers on insects included Henry Fleming, Rosemary Kenedy, and especially Jocelyn Crane. Miss Crane described one of the major insect projects at Simla in her *National Geographic* article "Keeping House for Tropical Butterflies."[4] Here she told of her observations and experiments with various species kept in large screened enclosures built for the purpose near the headquarters. She did not, however, abandon her work on fiddler crabs. After taking a trip in the mid-fifties to study new forms of this group in various parts of the world, she published a preliminary report on her findings in 1957. The next year she established four local species of fiddlers in captivity at Simla for additional studies. During these years Miss Crane's duties included effective directorship of the Department of Tropical Research, William Beebe having resigned on July 29, 1952, his seventy-fifth birthday, to become Director Emeritus.

In the latter part of 1955 Beebe went on his last major expedition, a trip of 144 days which took him over much of the route he and Mary had traversed in 1910: New York to Naples, then to Port Said, and on to India, Ceylon, and finally Singapore. Forty years earlier he had written of his fears for the survival of certain pheasant species threatened by changing habitats or relentless market hunting. Now he sought to discover their current status, although as an American he could not extend his itinerary to include the new People's Republic of China. He also attempted to study the birds of the Singapore area in order to compare them with those of Trinidad. However, neither a technical paper nor any other kind of lengthy publication grew out of this second long journey to the East. In subsequent years other writing projects were mentioned—a popular book on Simla, a detailed comparison of the birds of the temperate zones and the tropics; these also failed to appear. Only one extensive piece of writing for the popular audience came out of the Simla period, and it had little or nothing to do with the tropics. This was *Unseen Life of New York As a Naturalist Sees It,*[5] Beebe's last book.

In length—only 165 pages—and aspiration *Unseen Life* is clearly a modest effort; yet within its small scope it is in no way inferior to Beebe's major works. How his native New York once was, how it has changed, how it is now—this is broadly the subject of the book. Playfully but with an underlying intellectual seriousness, Beebe takes the reader on a journey through the geological ages and the biological phyla, ancient and modern, making easy, almost casual use of a lifetime of nature work at the desk or in the field. He is reflective and reminiscent but never morose, incisive without being mordant, sturdy and undismayed and even cheery, for all the unnatural shocks New York has sustained since the Indians sold it. Beebe knows that there has been destruction, even devastation, and especially in recent times, gross pollution of the environment; he also knows that nature has her own ways of adapting, and must finally have the last word, inexorable and unanswerable.

Unseen Life differs from other Beebe books in lacking a narrative base; it is not so much a personal tale of adventure as it is a series of descriptive and expository and speculative essays. Hence, perhaps, it is less vivid in phrasing, less charged with eagerness and tension. But Beebe is not merely going through the motions, turning out one more book for the marketplace or the shelf where twenty-one of his works already stood.[6] As he had done all his life, he was setting down his thoughts, communicating his enthusiasms, writing with unflagging spirit the last chapters of a remarkable man's essential autobiography. In the first of his books he had celebrated "the joys of very life," and here in the last he hailed our "manifest destiny of joy and awe in this life" on earth. In half a century many things had changed, but not William Beebe's ebullient response to the living world.

To fear that Beebe might at last fall prey to an autumnal kind of sentimentality is radically to misapprehend him—and also to forget that for him Henry David Thoreau was from first to last a challenge and guide. As Beebe wrote late in life, when you read *Walden* and then reread it, you "become less and less articulate but more quietly, completely satisfied."[7] The fearless rigor of the man, the cool unwavering eye, disquieting to some, Beebe found both sobering and inspiriting. A decade before Darwin, Thoreau spoke in the plainest terms of the power which nature holds over all the species, our own not excepted: "We go on dating from Cold Fridays and Great Snows; but a little colder Friday, or greater snow would put a period to man's existence on the globe." Now a century later we have William Beebe in a two-page conclusion to his final book, conducting us with impish high spirits to the frigid engulfment of human civilization in the present temperate zones, the consequent shrinking of seas around the world, and the arrival in New York of the beasts of the snow regions; and here is his last sentence: "Another ice age is on its way!"

Beebe once wrote of the deep thrill of seeing certain things in nature for the first time—a wild monkey, the rings of Saturn, a volcano in eruption; and he concluded: "When my time comes, if I could choose my mode of death, it would assuredly be of sudden heart failure from some unexpected, unpredictable sight such as one of these."[8] He was not so favored; instead he met his death not from sudden failure of the heart but from slowly failing health. His journey to Singapore at seventy-eight was a valiant effort not to be repeated. In the years following he traveled from New York to Simla when winter came, returning in the spring. Then in 1959 he left Simla for New York in May but went back to the tropics in October, establishing a pattern for avoiding cold weather which he followed to the end. Finally in 1962 Beebe spent the first two weeks of May in New York, but found the city too cold and took refuge again in Trinidad, perhaps knowing that he had but a short time to live.

Upon his death the *New York Times* stated that Beebe "had been ill for three years." Other sources also have suggested extended and even disabling illness, including near-immobility and loss of speech. But the physician who attended him through his last several years at Simla, the Canadian A. E. Hill, tells of a man robust in spirit even as the years encroached upon his body and claimed it at last. In particular Dr. Hill denies the more extravagant statements about Beebe's loss of physical powers. He writes:

> The quotation from Fairfield Osborn, stating that Will's final illness made him "virtually immobile and incapable of speech" goes much too far. Certainly he gradually weakened during the last few months of his life, and he was forced to spend increasing periods in bed, but he was able to move about the rooms of his beloved "Simla" virtually without assistance, almost up to his last day. Over his last year or so, he developed a recurring tendency to moderate slurring of speech, which made me suspect he had suffered a small stroke—perhaps more than one. Yet at times the impediment cleared almost entirely. It tended to worsen under tension. Thus, in the company of persons who annoyed him, or simply bored him, the slurring became more noticeable; as he was self-conscious about it, he would become increasingly taciturn. This undoubtedly gave some of his visitors the impression that he was, at times, more or less "incapable of speech."[9]

Clearly Beebe was distressed by his speech problem, and yet in sympathetic company he could make light of it. A close acquaintance reports that "mango-mouth" was Beebe's facetious term for the difficulty. But, as Dr. Hill states, this trouble occurred not continually but sporadically, with much depending on mood. Others who knew him give much the same testimony. One tells of a visit to Simla in 1961, during which Beebe treated his guest to a lengthy discussion of *A Monograph of the Pheasants,* turning the pages of these great volumes and commenting on many aspects of the work. Another relates how Beebe put an abashed woman visitor at her ease by joking about one of the antic hats he was in the habit of wearing; and a third, a woman friend of long standing, recalls that only a few months before his death Beebe challenged her to identify a statue recently placed on the terrace at Simla, and was delighted when she said, "Winnie the Pooh." William Conway, now General Director of the New York Zoological Park, tells how Beebe at age eighty-two filled his mind one evening at Simla with stories of vampire bats—and then at 2 A.M. sneaked in and pinched Conway's toe in imitation of a vampire bite, causing him to wake up with a wild yell of fright. And while none of this suggests an illness of three years, perhaps the most unequivocal statement on Beebe's alleged disablement came from a black Trinidadian who served him for many years: "No man—that is not true! And I was with him to the last!"

But the slow decline Beebe suffered in physical strength if not in spirits is undeniable. Again in the apt and forthright words of A. E. Hill:

> The illness which did most to restrict his activities over his last three years was a gradual cardiovascular insufficiency, complicated by the inexorable processes of simple old age. Towards the end, we talked several times of the possibility he might be better cared for in a hospital or a nursing-home in Port of Spain, but he would not hear of it. He was fanatically attached to "Simla," and certainly he could not have had more devoted care anywhere than he received there, under the constant supervision of Dr. Jocelyn Crane. "The Old Man's Friend"—pneumonia—which finally tipped the scales against him, was a development of his last few hours. It was not a fulminant or extensive infection, and would probably have presented little challenge to him if his general state had not been so weakened. My wife and I visited him, at his request, only two or three hours before his death, and still then he was the bright, cheerful and lovable person we knew so well.[10]

William Beebe died June 4, 1962, at Simla. Jocelyn Crane informed the authorities and Dr. Hill signed a certification of death by lobar pneumonia and senility. Funeral services were held at All Saints Anglican Church on June 7 in Port of Spain; burial took place in Mucurapo cemetery in that city. The headstone chosen was a rounded grayish boulder bearing a small bronze plaque: WILLIAM BEEBE 1877–1962. The grave faces west toward another with a tall gravestone that reads "In Loving Memory of LIOW NYAT EN" along with five vertical lines of calligraphy. Just to the northeast is the family plot of Sookram and Singh, and a little farther on, the grave of one Benjamin Preddie, his place marked by a sign hand-lettered on black tin. The visitor may find that Beebe's immediate plot is neatly scythed, but all around is the irrepressible tropical dishevelment of this most ecumenical and least thanatotic of graveyards. It scarcely needs said that Beebe himself had chosen this place to lie.

Scientific activity at Simla continued for another decade, reaching a peak in the middle sixties and then declining. In 1965 the Zoological Society and Rockefeller University established an Institute for Research in Animal Behavior, including within its structure the old Department of Tropical Research. Donald R. Griffin was appointed Director of the institute, and in December of that year he and Jocelyn Crane were married. Simla, renamed the

William Beebe Tropical Research Station, no longer had a central role to play, and it was becoming expensive to maintain. Eventually it was put up for sale, and by the spring of 1971 it had been virtually abandoned to the yellow-tailed cornbirds, the bananaquits and the black and coral heliconid butterflies—perhaps the same *Heliconus melpomene* whose lazy, insouciant, self-confident flight Beebe had enjoyed in Venezuela many years before. Although some hope was expressed that the Trinidad Field Naturalists' Club or even the island government might take over the installation to continue its work, nothing came of this, and in early summer Simla was declared closed.[11]

Elsewhere within the Zoological Society other changes were taking place. Beebe's "football field" flying cage for birds had become structurally unsound and the new bird houses of his early career were being supplanted at the Bronx Zoo by the World of Birds, a gray free-form building complex opened to the public in June 1972. The bathysphere, symbol of another major phase in Beebe's life, had survived wartime depth charge experiments conducted by the Navy and had been featured in an exhibit at the New York Zoological Society Aquarium in 1962; but marine studies had taken a new direction with the establishment of the Osborn Laboratories of Marine Science in 1967. Here as in all other branches of natural science, new times called for new forms and ways.

For something over half a century William Beebe had functioned in the Zoological Society as a dynamic and even dominant force. From his first innovations in the Department of Birds through his pioneer ventures in the ocean depths and his multitudinous researches in the tropical jungle, he had gone his own way, followed his own bent. As a public figure he was unmatched within the Society, and as the director of major scientific enterprises he was deferred to if not always unchallenged. He was Beebe; he put his stamp on things. But it is in the nature of such strong personalities that their ways of operating, their projects, methods, idiosyncrasies, do not long outlast the men themselves within a given structure. So it was with Beebe. His work had been done; it was not now to be repeated, but judged for its worth and place, and for its progeny.

:10 Man of Science

> Many of us can never hope to reach the clear heights of quick dynamic thought, and the genius of generalization which in the last analysis is the only *raison d'être* of facts and the search for facts. Most of us must be content to gather the bricks and beams and tiles in readiness for the great architect who shall use them, making them fulfill their destiny if only in rejection.
>
> "The Hills" (1916)

Whether William Beebe would achieve the clear heights or remain instead a gatherer of building materials for some other, more gifted seeker was still in doubt when this was written.[1] By "quick dynamic thought" and "genius of generalization" Beebe meant theoretical work, the intellectual structuring of observed facts in ways that demand that we observe those facts anew, perceiving them as related in novel patterns and implicit with unsuspected meanings. Doubtless Beebe was thinking first of Darwinism, foremost among all modern biological theories. He may also have had in mind certain theoretical problems taken up in times past by ornithologists—Mark Catesby speculating on the autumnal disappearance of chimney swifts, Buffon pontificating on the origins of New World birds, Benjamin Barton (and many before and since) mulling over questions of bird migration, Wilson reflecting upon various bird superstitions—or, in more recent times, many of his fellow bird men, including even Theodore Roosevelt, arguing over the theory of protective coloration. And in fact Beebe by this time had submitted two of his own scientific formulations to the scrutiny of his peers. For all the seeming modesty of the passage quoted, Beebe surely hoped that his work would prove to be more than mere bricks and tiles.

This was the period of his intensive engagement with the pheasant enterprise, and Beebe's first synthesis of data emerged from his search for a new classification system for the numerous genera of this group of birds. As he reported in *Zoologica* in 1914, his study of the problems had shown him "how artificial and arbitrary [had] been the previous attempts to find some character which would be of use in separating the major divisions of the family," and as a consequence he "made elaborate tabulations of several scores of what seemed characters of significance" during his lengthy study of pheasant collections in various museums. Of these "characters" only one proved at last to be truly diagnostic—"the *mode of moult of the rectrices*" or tail feathers. Briefly, Beebe discovered that his nineteen genera fell naturally into four sub-families, depending on whether tail feathers were moulted from the central pair outward to the edge, or the reverse, or proceeded in both directions from the third pair, or the fifth. Though he stated that "any classification must finally be the result of a balanced weighing of . . . varied characters," and declared himself willing to abandon tomorrow his tail moult theory if "a more consistent, logical factor be discovered," nevertheless he viewed his classification scheme as an important contribution.

Then in 1915 in *Zoologica* Beebe published "A Tetrapteryx Stage in Ancestry of Birds," a theoretical paper offering a conjectural creature, the Tetrapteryx, as an early progenitor of modern birds. In contrast, for example, to the relative abundance of fossils of ancient reptiles (including flying varieties), there is no fossil evidence at all of birdlike creatures prior to the appearance in Jurassic times of *Archeopteryx lithographica.* Excellent impressions of this species were discovered in a Bavarian quarry in 1861 and 1877, and a less complete one in 1956. They show a crow-sized primitive bird which had departed so far from its presumed reptilian ancestry as to have acquired an ample feathered tail, and forelimbs already evolved into broad and fully feathered wings. It is an article of evolutionary faith that nothing so specialized as Archeopteryx could have sprung from reptile progenitors without a long period of development through intermediate forms. But no such forms are known from the fossil record; they must be hypothetically reconstructed.

Beebe named his reconstruction Tetrapteryx, the four-winged one. As he proposed it, this animal was more reptilian than was Archeopteryx, but had already developed enough feathers to make the forelimbs winglike. The hind limbs featured the major novelty upon which Beebe's theory was based: a fringe of feathers projecting backward from the thighs which in effect converted them into small, rounded "pelvic wings" to aid Tetrapteryx in gliding through the air.

The major proof for the existence of such wings came from Beebe's discovery of incipient quill feathers growing on the thighs of a white-winged dove four days after hatching. He found similar feather growth on the thighs of a jacana embryo. Particularly in the case of the dove, he thought of these quills as atavistic and hence clear evidence that pelvic wings had appeared in the line of bird ancestry prior to the Jurassic period. It was assumed that these rear wings became less functional and disappeared with time, although Beebe said he found vestiges of them in the 1877 specimen of Archeopteryx in Berlin. He declared in summation:

> Millions of years after they were of use, the feathers of the pelvic wing are still reproduced in embryo and nestling. And for some unknown reason, Nature makes each squab pass through this tetrapteryx stage. . . . No fossil bird of ages prior to Archeopteryx may come to light, but the memory of Tetrapteryx lingers in every dove-cote.

Another important area of theory for Beebe concerned the development of living creatures in the Galapagos archipelago. His early adherence to the land-bridge theory and later acceptance of counter-evidence derived from his own observations have already been discussed. Here as elsewhere Beebe declared himself ready to relinquish his ideas if better ones could be shown him. But he took his work seriously, and in one passage in *Galapagos: World's End* he listed his pheasant classification system, his tetrapteryx theory, and his views of Galapagos evolution as foremost among his theoretical efforts.

As noted, it took Beebe himself only two years to discover and cheerfully admit serious flaws in the last of these. About the same time the Danish ornithologist Gerhard Heilmann was taking up the tetrapteryx hypothesis, discussing it at considerable length in *The Origin of Birds* (1926).[2] Heilmann "with some excitement" began his scrutiny of Beebe's theory by studying the nestlings in the Zoological Museum collection in Copenhagen, but with wholly negative results: "there was not in any of [these nestlings] the slightest trace of a 'pelvic wing.'" He then turned to the species lowest on the evolutionary scale, the ostrich and emu and others, in the hope that such traces might show up in these, the most primitive in structure of living birds. Yet here too "there was not the slightest trace of sprouting wings on the hindleg" of the young birds. Nor were any such unusual structures found among chicks of many other bird families investigated by Heilmann. Even his study of pigeon nestlings, closely allied to those of Beebe's white-winged dove, afforded Heilmann no proof. Instead of Beebe's incipient "pelvic wing," he found on the thighs of squabs "a series of permanent feathers, and no atavism. If it were a genuine relic from such a very remote past, it would make its appearance, like a glimpse, in the embryo or squab, quickly to vanish again."

Heilmann went on to discuss the anatomical and evolutionary implications of wings growing on a creature's hind limbs, finding such a growth a threat to survival, and hence not likely to have persisted in the line of bird ancestry, even admitting that it had in fact once developed. Hind wings would deny the kind of free movement necessary in the struggle for existence, and presumably bring the tetrapteryx line to an end. (By way of contrast, Heilmann's own hypothetical "Proavis" exhibited no such disabilities.) In effect, then, Heilmann rejected every important aspect of Beebe's theory, and ornithological literature since that time has accorded it little notice. Heilmann, on the other hand, is still cited as an authority.

A full appraisal of Beebe's system of classification of the pheasants did not appear until Jean Delacour brought out *The Pheasants of the World* in 1951.[3] However, in 1934 James Lee Peters issued the second volume of his *Check-List of Birds of the World,* in which he took up the pheasants, acknowledging help from other scientists on a few of the genera but not mentioning Beebe, even while radically revising his classification. Beebe had offered 19 genera, 63 species and 49 subspecies; in the Peters check-list the equivalent figures were 21, 50, and 106. As Beebe himself remarked in the foreword he wrote for Delacour's book, the ink was scarcely dry on *A Monograph of the Pheasants* before new discoveries began to render it obsolete. To this the Peters volume gave initial confirmation.

Delacour, working as Beebe's good friend and professional associate, found it appropriate to discuss his own classification at some length. In certain ways Beebe's brief foreword and Delacour's discussion are exercises in polite mutual equivocation. Beebe, acknowledging his work to be outdated in many respects, professes delight that Delacour has chosen to bring out the current work to supplement—but not supplant—the field work of the monograph, especially through observations of captive birds. Beebe notes that technical classification of the pheasants has greatly changed, but adds rather lamely that "the common terminology remains unaltered and forms an unbroken link with the many references in ornithological literature." Delacour for his part states that he began his work "where Beebe left off," and speaks of *A Monograph of the Pheasants* as "a veritable monument to the glory of the birds, contributed by a naturalist with great literary talent" and vast field experience. His own work encompasses "new dimensions" and "fresh information," and therefore "completes and brings up to date Beebe's *Monograph*."

Shortly, however, Delacour launches into a section called "Systematics," and essentially repudiates the method of classification Beebe had devised and later claimed as a major accomplishment. He offers Beebe's tail-moult theory the lefthanded compliment that it "coincides generally with other features and corroborates specific and generic affinities already otherwise indicated," but in contrast to Beebe he specifies other important criteria for judging taxonomic relationships, including both a variety of natural habits or attributes and studies in hybridization, cytology, and serology. In particular he warns against "any . . . system based on unique characteristics," and concludes by insisting that all aspects must be taken into account "to form a reasonable appreciation of true relationships."

Delacour's classification resulted in 16 genera, 49 species, and 122 subspecies. Obviously the number of known, recognizably distinct pheasant forms had greatly increased since Beebe had done his work, although both Delacour and Peters had somewhat reduced the list of full species while adding greatly to the subspecies. Delacour acknowledged that Beebe had recognized 46 out of the 49 species admitted to the new list, but this did not prevent Delacour (or Peters before him) from demoting Beebe's major new species, *Ithaginis kuseri,* to *Ithaginis cruentus kuseri,* one of a dozen subspecies of the Himalayan blood pheasant.

If Beebe thus had little success with his theoretical work in birds, he was no more notable for his efforts at synthesis in other areas of natural history where his professional training was not so extensive. It has been indicated that Beebe's reception among ichthyologists was less than warm. In entomology, another important concern, Beebe worked with devotion and persistence without being recognized as a major figure within the discipline. To define him as a generalist or an eclectic and thereby dismiss him is not enough. William Beebe was a man of remarkable intellect, great powers of observation, and a splendid capacity for response. It is patent that he wished to combine his gifts to achieve "the clear heights," even possibly to become "the great architect" of which he speaks; but essentially, whether in ornithology or elsewhere, he failed.[4]

What is absent or deficient is part at least of failure; and so it was with Beebe as an aspiring theorist. His talents are plain enough, but his deficiencies are evident also. This man who laid claim to a Bachelor of Science degree from a major institution in fact had none. He had never submitted himself to the discipline and drudgery of an undergraduate program, nor to doctoral work at a more demanding level of intellectual commitment. In areas of science already grown quite complex—geology, paleontology, evolutionary theory, embryology—he lacked formal training, and in the course of a crowded career he had little chance to acquire competence in such studies, even as they burgeoned and their complexities increased.

Lee S. Crandall says it was on Henry Fairfield Osborn's advice that Beebe left Columbia to go to work at the zoo, because a year there would have more value than one at the university. Strange advice from a professor, perhaps; but Osborn may have understood his young friend Beebe pretty well, recognizing both his venturesome qualities and his restiveness under academic discipline. Temperament, in short, precluded the classroom and laboratory training which might have made Beebe another kind of scientist. At the same time it embraced a self-confidence occasionally almost Olympian, as in the undaunted leap across 200 million years from a squab's thigh feathers to Tetrapteryx, or in the statement Beebe gave to the *New York Times* in 1926 that the "wash boiler" depth vehicle would need no lights because the sea "at the depth of a mile or so is highly illuminated by the luminous organs of deep-sea fish."

But temperament also produced the man we know, the indefatigable naturalist. It is bootless to ask for another person and another career, and useful to recall that science is served by more than theory. In a score of books and a hundred *Zoologica* articles Beebe had offered a lifetime record of scientific work, most substantial bricks, beams, and tiles if not some kind of splendid edifice. At the most practical level were the two field works on fishes, a book for Bermuda, and a set of articles for Haiti; there was *The Bird, Its Form and Function,* as close as Beebe came to writing a textbook; and there was *Tropical Wild Life,* a group of relatively technical papers collected into a single big volume. And of course there were the many volumes of scientific adventures, and the pheasant books, part technical treatise, part field report, and part romance.

Surely these too served science, but in just what fashion is not always easy to define. In the study of insects, for example, which deeply engaged Beebe's interest both early and late, his essays ranged from the most concentrated, as in several ant studies, to the broadest and most speculative, as in "The Bay of Butterflies"—but what is the professional entomologist to make of them? The larger number are accounts of absorbed observation of the living world, not detailed life histories or laboratory experiments on captive insects or museum correlations of dead ones. William Morton Wheeler, a sensitive and graceful writer as well as a noted entomologist, posed the problem engagingly when he contrasted professional researchers "obsessed by problems" with more fortunate "amateur entomologists, who have not been damned professors"—the former doomed in Hades to sit forever identifying specimens, the latter free to pursue "gorgeous, ghostly butterflies until the end of time."[5] In short, there is, or should be more to the study of living creatures than narrow problem-solving research; the naturalist has his place along with the specialist. John C. Pallister, senior entomologist at the American Museum of Natural History, makes much the same point: "Beebe did original work in entomology, but he was more of a naturalist than anything else, and many Ph.D.s might have regarded him as a crank. But he may have regarded them as cranks too."[6]

In marine sciences the problem is more complex, involving as it does Beebe's true pioneering in the study of creatures at great depths, and his novel field work in shallow water. Using his helmet, for example, Beebe was able in a few days in May 1925 to observe twenty-three out of the thirty-eight known shore fishes of Cocos Island—and also to identify fifty-seven other species never before recorded at Cocos, along with about thirty more seen too briefly to be named. Nor was this particularly unusual for him. Beebe made a great number of observations of marine organisms in their own environment; he discovered new species, whether by direct encounter or by more conventional methods of collecting; he speculated on many aspects of life in the sea, though he offered no important theory. In the ongoing quest that marks all science, it is proper that books and monographs and technical papers should appear, be duly noted and judged, and, with few exceptions, fade with time. So with Beebe's marine work—but his analogy of the Martian visitor still applies. Whether or not Beebe was regarded by marine scientists as a notable figure, his work in both helmet and bathysphere nevertheless afforded new directions for study, some of them yet to be fully exploited.

With the coming of the snorkeler and scuba diver, fish watching by a host of amateurs has impinged willy-nilly upon the science of ichthyology, with professional reactions doubtless ranging from delight to abhorrence. The same has been true in ornithology for nearly a century, at least from the time bird watchers came together in the Audubon societies, the first of which was organized in 1886. Among all the life sciences, none is so pervaded by general awareness as the study of birds, none is so "public"; and among all American biologists the name of John James Audubon is the best known, his work the most widely appreciated and displayed. Hence the notion of bird science as recondite, arcane, "pure," has been frustrated from the start; public interest has been too active, and indeed public knowledge has been too extensive to keep the discipline within professionally esoteric bounds. The staff of the Museum of Comparative Zoology at Harvard once included both James Lee Peters, America's foremost avian taxonomist, and Ludlow Griscom, most expert of our ornithologists in field identification; and it is perhaps symbolic that Roger Tory Peterson, presently our most acclaimed bird scientist, used the vast skin collection of the one and the matchless outdoor experience of the other in researching the first of his famous field guides to birds.* So Beebe was not the poorer ornithologist for being a popular one, any more than was Frank M. Chapman for being a tireless propagandist in the cause of amateur bird study and conservation. And if Beebe's theoretical work came to little, his close studies of individual species—the hoatzin, the tinamou, the toucan, the bat falcon—were and are prime sources of information, just as his intensive field work with the pheasants stands unrivaled

* Lest Mr. Peters sound as dry here as one of his own bird skins, permit a personal recollection. A peregrine falcon once came to spend the winter on a tower near the museum, and when it was pointed out to him through a museum window, Mr. Peters, his eyes sparkling, said: "Splendid bird—splendid!"

still. And again, because bird study has a following at so many levels of understanding and attainment and response, there are those readers whom the least scientific of Beebe's bird work may still charm and enlighten.

But for a time in this century it appeared that life science generally would relinquish its concern with natural life for a new (or at least newly modish) concept of science. Persons trained to study not living organisms but lifeless particles of matter—molecules and their smaller components—entered and soon appeared to dominate the field. It is a commonplace that some of our best young intellects were attracted to the study of mathematics, physics, and chemistry from the time of the Second World War onward. Such people inherited and carried forward a record of stunning practical achievement, whether lethal or life-enhancing, and they worked within a framework of theory so precise and elegantly articulated that their potentialities seemed boundless in whatever direction they chose to turn. And many—some perhaps in reaction against the baleful import of much that physical science and technology had already spawned—turned to life science, bringing with them the confidence of their exactitudes and the precision tools, both technical and intellectual, with which to create a new biology.

It is suggestive that the most spectacular recent discovery, perhaps indeed one of the most germinal of all biological formulations, should have been the joint work of a young American biologist, James D. Watson, and a British physicist twelve years his senior, Francis Crick. Their odd contrivance, a molecular model of spiral-staircase design which they completed in 1953, soon became known to the lay public by the name Double Helix, or perhaps DNA, for the deoxyribonucleic acid molecule it represented. The public also knew that a Nobel Prize was awarded to the two discoverers; beyond that, this product of the new biology, unveiling, as was claimed, the very secret of life, remained as remote from common understanding as the more arcane disclosures of modern physics. And indeed molecular biology shared with physics common assumptions and techniques and a vast potential for extension within its avowed limits. Genetics, embryology, endocrinology, physiology, and psychology were both contributors to and beneficiaries of the new synthesis; eugenics, that ambiguous Doppelgänger of genetics, acquired novel and potent instruments; and with increasingly sophisticated and exact understanding of thought processes has grown the capacity for their manipulation and control. Those who came from the physical sciences to biology to find refuge from moral unease must again confront the question posed some nineteen centuries ago: "But who controls the controllers themselves?"[7]

Questions considerably more mundane, however, troubled those life scientists who could not or would not share the new mode. As one historian put it:

> Molecular biology is Cartesian biology . . . based on the premise that biological mechanisms can be reduced to inanimate chemical and physical processes. In mid 20th century, biology is dominated by chemistry and physics. Difficulties arise, however, when attempts are made to account for the entire organism on the basis of molecular biology. Genetic coding, which represents one of the most important discoveries of the present century, will not account for all embryological and evolutionary changes among living things. If all living processes could be reduced to chemistry and physics, there would eventually be no need for biology.[8]

In other words, it appears that the old disciplines are fast becoming obsolete, and their practitioners with them. Under the new dispensation, where is there a place for a botanist or an entomologist? Who wants a limnologist, a dendrologist, even an ecologist? What of the ornithologist—has he any longer a function within the structure of research teams or museums or universities? And of all such people, the naturalist is surely the oddest, a figure not merely old-fashioned but antique. A naturalist studies nothing less than the living world, perhaps out of love for it; and for such a man or woman, where is there a meaningful place?

The word "woman" suggests a wholly persuasive answer. In 1962, in the midst of the putative triumph of the new biology, Rachel Carson brought out *Silent Spring.* This book without question influenced thought and affected practice in a way analogous to important discoveries from the laboratory; and it was written by a naturalist, albeit one both famous and exceedingly learned. Miss Carson had spent many years with the U. S. Fish and Wildlife Service as a biologist and editor, publishing during that time *Under the Sea-Wind* (1941) and *The Sea Around Us* (1951), and later *The Edge of the Sea* (1955). Now she has ranged more widely and written more urgently; the force behind *Silent Spring* was her deepening disquiet over the state of the environment. She saw the old dangers worsening even as new ones appeared; and like Beebe, she had the true naturalist's sense that numberless threads make up the earth's environmental fabric, enfolding not only life but non-life, the animate and inanimate as well. Man from the beginning of his self-awareness had strained and rent that fabric, but gradually, over centuries and millennia of time; only recently had he brought to his war against nature new weapons of mass destruction, in particular chemical pesticides. Famous victories ensued, with final triumph exultantly foreseen; but Rachel Carson was among the first to perceive such victories to be hollow and even Pyrrhic, and such triumph a prideful delusion. Four and one half years of research, travel, and correspondence with other trained observers went into this final book. Many hailed the work and resolved to heed its message; others attacked it as exaggerated, shrill, overwrought; and there were those who appraised *Silent Spring* not for its accuracy or worth, but for the vital threat it posed to their freedom to go on profiting from devastation.

Rachel Carson, born in 1907, had no evident desire to participate in the new biology, nor was she suited to it temperamentally or by special training, although she had done graduate work in genetics. If molecular studies symbolized a break with the past and a radical narrowing of areas of biological concern, her work represented continuity, the persistence of involvement with that broadest and most complex of systems, the world of life. In point of time she was the next generation from William Beebe, and intellectually his offspring too, acknowledging him as both friend and inspiration. There were many others, men and women who had grown up knowing Beebe through his work, his books, his personality, and not least, perhaps, his fame.[9] If they specialized, they did so among the great orders of living beings, and if they followed Beebe, even as specialists they retained a strong sense of nature as a great web of being, instinct with beauty and even mystery. To the degree that they dealt with forms of life interacting within the larger natural environment, their trend was ecological, as Beebe's had been from the start; and to the degree that they cherished this abounding world, they were conservationists—and at the level attained by Rachel Carson, splendid propagandists for life.

In the light of things said earlier about Beebe's concern with conservation, perhaps a summary is all that is needed. In 1906 Beebe was already speaking of "the most terrible of Nature's enemies—man,"[10] pointing to "dangers,—new and wide-spread,—

which are now affecting the environment of every creature of this world," and hoping for better things: "May the naturalists of to-day realize their opportunity and do their best to preserve to us and to posterity what is left to us of wild life. If not, let us pity the Nature-lover of two hundred years hence!"[11] In 1909 he praised the Indians of South America for not wasting "powder, shot or arrows on so-called sport," and hence not endangering the hoatzin. "Until the 'civilized' tourist penetrates to these regions, the Hoatzins are safe. . . . So helpless are they, that, given a week's time and a shot-gun, one man could completely exterminate them in the colony of British Guiana."[12] In Burma in 1910 Beebe was charmed by the great number of birds, ascribing it to the Buddhist belief in the sacredness of all life, while Christian lands must depend on bird protection societies just to keep down the slaughter. On the Galapagos he was poignantly aware of the vulnerability of many native forms of life, entreating at one point that such creatures be permitted to go on sharing with human beings our common earth.* Beebe's sea books not only spoke of endangered marine species, but recorded increasing pollution of the waters themselves.

Although his jungle books dealt with areas largely unspoiled and creatures not yet in peril, Beebe often noted how easily things might be altered by human intervention. In *High Jungle,* however, he offered both the depths of night and the depths of the sea as areas of nature which might still resist the remorseless advance of destruction. In *Unseen Life of New York* he anticipated by some years the current complaint against "the atmospheric pollution of automobiles and the steady closing in of lofty buildings"[13] threatening what remains of life in our cities. While he took sardonic comfort from nature's abiding power to speak the last word, Beebe nevertheless remained a conservationist to the end, even to the extent of ruling against the taking of specimens at Simla—"Observe, but do not collect."[14]

Conservation in earlier times tended to stress particular dangers faced by individual species: plume birds threatened by the millinery trade, bison by relentless hunting, redwoods by unchecked lumbering, the whooping crane and ivory-billed woodpecker and

* It is perhaps fortunate that Beebe did not live to read Roger Tory Peterson's absorbing article "The Galapagos: Eerie Cradle of New Species" (*National Geographic,* April 1967) with its dismal tale of the fate of the land iguanas Beebe had enjoyed studying on Seymour: American servicemen stationed there, bored but amply supplied with bullets, had used the great reptiles for target practice, exterminating them.

California condor by the disappearance or disruption of needed habitat. More recent threats to many species are of quite a different kind. Although the roseate tern is no longer killed for human adornment, and indeed has taken refuge in guarded nesting areas, it has lately begun producing chicks with a bizarre assortment of deathly deformities, from crossed bills to four legs.[15] Birds even better known, such protected species as the bald eagle and the peregrine falcon and the osprey, appear less and less able to produce fertile eggs or bring them to hatching. Laws and sanctuaries alone no longer answer the need. These creatures are imperiled not by gunfire or encroachment but by forces hidden within the environment; in short, their problems are ecological. This kind of understanding Rachel Carson brought to *Silent Spring,* and William Beebe prefigured by his lifelong concern with natural interrelationships.

In 1909, half a century before the term came into common use, "Ecology of the Hoatzin" was the title Beebe gave to a twenty-page article in *Zoologica.* Here and in later works on this bird, he had chosen a most illustrative example to show how one creature fits into its larger environment. Many aspects of the hoatzin suggest fatal vulnerability—its large size and loud voice and indifference to concealment, its apparent stupidity and awkward flight, its habit of nesting in highly visible and accessible colonies, the narrow limits of its diet, and the markedly restricted nature of its habitat. Yet the very fact of its survival as a locally abundant species proclaimed the hoatzin a success. Beebe's field research on the hoatzin threw much light on the ecological patterns within which this remarkable bird had been able to persist and modestly prosper.

The physical scale of Beebe's ecological studies ranged very widely. He recommended in *The Log of the Sun* that the investigator choose "twenty square feet of snow" and with "chin upon mittens and mittens upon the crust" he will find that "the eye opens upon a new world,"[16] while the hand lens reveals even more complex and detailed aspects and relationships than the naked eye can perceive. In "Fauna of Four Square Feet of Jungle Debris," a *Zoologica* paper of 1916, Beebe radically altered the latitude of his study area but not his devotion to minute analysis and tabulation. In the fifth chapter of *Tropical Wild Life* he broadened his scope to embrace a thousand acres of cutover jungle near Kalacoon, part of it planted in rubber trees, the rest already reverting to second growth. Beebe's "brief résumé of the general ecology" included a relatively detailed account of what happens when a clearing is made in mature jungle, with particular stress on the appearance and growth of new plant and animal forms.

In keeping with the diversity of its subject matter, *Nonsuch: Land of Water* contained several close studies of Bermuda life which demonstrated the links between land and sea and the shore that marks their mingling. In other writings Beebe's range grew even wider, but he never lost his early love for intensive study of interrelated life. In the passage already quoted from *High Jungle* he reaffirmed the "steady concentration on a variety of living organisms and their constant comparison" which is a central principle of ecological science; and one of the last of Beebe's important *Zoologica* papers was his report in 1952 on the ecology of the Arima valley in Trinidad, the setting for his final studies.

So it is clear that for many decades ecology was one of Beebe's major concerns, at least as he understood the term; what is not so clear is whether recent ecologists would be willing to admit Beebe to their company, except perhaps as a remote ancestor. Although William Morton Wheeler used "Ecology" in 1928 as a substitute word for natural history, the branches of ecological work have proliferated greatly since that time, and have tended to depart from outdoor studies and arrive by different paths in the laboratories or among the computers of the new biology.[17] Again, it seems, those of the naturalist persuasion are beset or at least far outnumbered. Yet ecology has always been interdisciplinary by its very nature, and hence not easy to delimit and define. Encompassed, then, within the most tolerant of definitions and the most catholic range of studies, perhaps there will remain beneath the rubric of ecology both the offspring of René Descartes, relishing their certitudes, and the sons and daughters of William Beebe who, even when the wind is north-north-west, at least know a hawk from a handsaw, and take much pleasure in the difference.

As early as 1919 Wheeler used the term "ethology," but not in a clear context. Webster's quarto dictionary nine years later defined the word as "bionomics," which in turn meant "The branch of biology which deals with the relationships of organisms to their environment; ecology." One of those who has rescued us from this state of semantic redundancy is Konrad Lorenz, the Viennese physician and biologist best known for *King Solomon's Ring* (1952) and *On Aggression* (1963). From the beginning of his career Lorenz has used ethology to designate "the scientific study of animal behaviour," and according to the British scientist W. H. Thorpe, he can be "rightly regarded as the founder of the new movement."[18] One may protest that neither the name nor the study is really new, but it is plain that Lorenz and such coworkers as Thorpe and Nikolaas Tinbergen have given ethology a new trend and definition.[19]

Readers of *King Solomon's Ring* are likely to recall most vividly Lorenz's experiments with young waterfowl, especially his "imprinting" of goslings and ducklings by using the natural parental call and thereby persuading them to accept him as surrogate parent. But Lorenz relates how initially he failed to discover the right note to attract ducklings hatched from wild mallard eggs in an incubator. Then it occurred to him that a muscovy duck had suffered the same rejection, while a white barnyard duck had the mallard ducklings following her immediately. It did not take long for Lorenz to realize that this latter bird was descended from mallards, and had retained the ancestral call notes throughout. When he gave the same calls the ducklings accepted him at once.

A similar use of sounds had fascinated William Beebe from the beginning of his study of nature. As with many another youngster growing up in North America, his earliest success came with the chickadee, that diminutive bird which comes so confidingly to any decent approximation of its *phe-be* whistle. In the tropics he had the same experience with a shy species of jungle wren, which swiftly crossed a quarter of a mile of swamp in response to a particular note, and with a young Galapagos mockingbird, which flew nearly in his face when he gave the right summons. Beebe seems almost an earlier incarnation of Konrad Lorenz when he imitates the adult tinamou and causes the tinamou chick to run toward him and crouch at his very feet. Later, when he put his face to the ground and called again, the chick ran all the way to Beebe's mouth. Much the same technique worked for adult sloths, who started slowly in Beebe's direction when he whistled a baby note pitched at a high E flat. Even a sphinx moth was prompted into defensive action when Beebe gave a high-pitched buzz simulating the ichneumon fly.

Another way Beebe studied the behavior of wild creatures was by participating as directly in their lives as his six-foot body would permit. He was always willing to enter the environment of a given species in the most intimate way possible, to crouch or crawl or slither, to submerge himself part way on a jungle beach or wholly in a tropic sea. When Lorenz goes snorkeling in the Florida keys observing patterns of aggression and defense among coral reef dwellers, he is doing as Beebe had done years earlier with water-glass and helmet, and for much the same purpose, to discover and understand the behavior of marine animals in their natural setting. And when Lorenz squats and waddles the better to play parent to ducklings, he vividly recalls Beebe and the albatross he joined in a courtship dance on Hood island in the Galapagos. After witnessing the astonishing ritual of bowing and posing and bill-crossing performed by an albatross pair, Beebe approached a lone bird and bowed low. To his delight, the great solemn creature bowed in turn, but when it approached to cross bills, Beebe could not, to his sorrow, reciprocate. In the same empathetic spirit he crawled on his belly among Galapagos iguanas and sea lions, studying their ways by sharing in their doings.

At its highest level, Beebe declares, this technique of research permits men to "enter into the very life feelings and intimate habits of wilderness folk"[20] and offers ethological insights which cannot otherwise be obtained. In the final paragraph of *Nonsuch: Land of Water* he sums it up this way:

> Before we can have the complete solution of the whys and wherefores of [certain kinds of behavior], there must be a great deal of uncomfortable climbing and diving, hiding in unpleasant places, getting wet and hot and cramped and weary. And then after we have tried to be sandpipers and ants, silversides and mackerel, we may attain to the honor of such knowledge as our prejudiced, but humbled minds will permit.[21]

In "Old-Time People" Beebe told of a man who caged himself in Africa to learn gorilla talk. Thoreau spoke of becoming neighbor to the birds at Walden, not by putting them in cages but by building his own cage near them. In establishing his research stations in the midst of tropical forests Beebe had given new impetus to the same idea, pursuing the study of a single area and its various species over a period of years. Although such field research was overshadowed by the postwar flowering of new branches of biological science, it nevertheless endured. For one reason, there was still a very great deal to know; behavioral studies, whether given the name ethology or psychology or even animal sociology, had far to travel in understanding how lives are lived in a state of nature. Much might be learned from captive or domesticated animals, but confinement affords at best an approximation of natural conditions, and, as in physics, the very act of observing under controlled circumstances may influence the reaction of what is being observed.

The man with the cage in gorilla country—an actual person named Garner—derived little enough in the way of vocabulary from his self-incarceration. Gorillas are in fact rather shy beasts, and the unique spectacle of a pale-faced fellow primate behind bars was calculated to limit contacts between *Gorilla gorilla gorilla* and *Homo sapiens* severely. (Beebe suggests that any ape words collected were probably scurrilous in any case.) Fear of the gorilla's alleged ferocity had prompted poor Garner to devise his cage, and fear of authentic human lethality had doubtless motivated his skittish subjects. But no great success had attended other, more conventional efforts either; up to the middle of this century neither race of this great African ape had been intensively studied in the wild. Then with the appearance in 1963 of George B. Schaller's *The Mountain Gorilla: Its Ecology and Behavior,* biological science acquired its first extended and authoritative treatment of the ways of this animal, and with the same author's *The Year of the Gorilla* (1964)[22] the interested public could share in the good fortune.

The Schallers, George and his wife Kay, had gone Thoreau and Beebe one better, not only taking a cabin in a forest clearing to be neighbors to wild creatures, but using it as their base camp over a period of a full year to study a single species, with many excursions in adjacent regions for further work. And of course there was no thought of self-protection in a cage of iron, or even with weapons; Mr. Schaller (and often his wife also) went unarmed into the forest, seeking out gorilla bands, slowly winning their trust—or perhaps merely their benign indifference—and observing their ways at times from a distance of a few yards.

The Schallers did most of their work in the mountain forests of Albert National Park on the eastern border of the Belgian Congo, leaving in the early fall of 1960 in the midst of the turbulence which came as colonial rule ended and the Congo was declared independent. That same summer, about 200 miles south along the eastern shore of Lake Tanganyika, another intensive study of anthropoid behavior was begun by the young English scientist Jane Goodall among the chimpanzees of the Gombe Stream reserve. In the course of several months of arduous work she too was required to gain the confidence of her subjects, often venturing far into the woodlands in search of a chimpanzee band, and sleeping where darkness overtook her. At least twice she was struck by male chimpanzees, more out of curiosity or exasperation than hostility. In contrast to the rather placid and introverted gorilla, the chimpanzee tends to be an extrovert, subject to moods of high excitement; but, as with gorillas, group life among chimpanzees showed little serious intraspecific violence.

There are many other similarities between these closely related apes, including permissive attitudes toward offspring and mating habits quite lacking in rivalry. But Goodall discovered two important traits of chimpanzees hitherto doubted or unsuspected, the use of tools and the killing and eating of other mammals. These apes regularly break off grass stems or small twigs and push them into termite mounds to snare their insect prey, and they will on occasion catch and eat monkeys, young antelopes, and bush pigs, and perhaps other warm-blooded creatures as well.

Presently Jane Goodall began publishing the results of her research either in scientific papers or popular articles and books. "My Life Among Wild Chimpanzees" came out in the August, 1963 issue of the *National Geographic Magazine,* describing the first two years of her work and offering an abundance of photographs (including fine pictures of chimpanzees using tools) taken by herself or Hugo van Lawick, the man she later married. Subsequently the National Geographic Society published the book *My Friends the Wild Chimpanzees* (1967) which brought her story up to date, and in 1971 she drew upon a decade of study for *In the Shadow of Man,*[23] a work absorbing in content, nicely candid in statement, and often beguiling in style.

George Schaller meanwhile had shifted his research to mammalian life in Tanzania's vast Serengeti National Park, concentrating on lions and, as with gorillas, commonly observing them at close range. After three years of intensive study he published "Life with the King of Beasts" in the *National Geographic Magazine* for April, 1969. Among the most striking of his discoveries was the fact that territorial fights among male lions can cause the death of one of the contestants and the eating of his now defenseless cubs by the victors—and, most strangely, by the returning mother also. In these violent and sanguinary acts the contrast between lions and gorillas is extreme; but just as Schaller's scientific field work with the apes made the old-fashioned gorilla hunt appear both fraudulent and cruel, so his courageous intimacy with lions made the celebrated perils of killing them from afar with high-powered rifles seem a pretentious sham. William Beebe had insisted that his motto "Watch 'em Alive"—which in *High Jungle* he posed against "Shoot 'em and Skin 'em" and "Bring 'em back Alive"—was not for sissies, and by their enterprise and courage George Schaller and Jane Goodall surely proved him right.

In many ways these young scientists are to the childless Beebe his intellectual descendants, his symbolic grandchildren. They are adventurous and highly articulate outdoor biologists, as he was, and they have worked with two of his foremost institutions. The New York Zoological Society under Fairfield Osborn sponsored the first gorilla trip for George Schaller; later he became a Research Zoologist for the society's Institute for Research in Animal Behavior. The National Geographic Society backed Jane Goodall in a similar fashion, and she was inspired and encouraged at the start of her career by the late Dr. Louis S. B. Leakey, the famous paleontologist whose work on ancient man at the Olduvai Gorge in Kenya was supported by Geographic Society research funds. Again like Beebe, both Goodall and Schaller bring to their studies a broad ecological outlook which calls for extensive knowledge of the environment in which their subject species live. George Schaller, for example, shows awareness and understanding of a wide range of life, and in fact took undergraduate degrees in both zoology and anthropology, and later worked in other fields, including bird behavior. Both he and Jane Goodall have far more academic training than Beebe received; Schaller earned his doctorate at the University of Wisconsin and has done advanced study elsewhere, while she was awarded a Ph.D. at Cambridge. Yet with all this, grace of language still attends them, and they reject that guild consciousness which decrees certain permissible outlets and audiences for published work while coolly reprehending all others—among the latter, the general public.

The scientist who seeks public as well as professional recognition necessarily proceeds at his own risk. Among his peers, that objectivity which properly informs scientific judgment may not always be applied to questions more subjective in nature, especially when fame or notoriety is thrown into the balance. It was one thing for Beebe to publish research papers in *Zoologica* or *Copeia* or *The Auk,* but quite another to be featured in a *New Yorker* cartoon, to appear in *Better Homes and Gardens* and the *Literary Digest* and *Saint Nicholas,* to broadcast his undersea doings over an international radio network, or to perform as guest expert on NBC's "Information, Please." Faced with such a range of activity, some of his fellow scientists found Beebe rather a strange one, an exotic difficult to classify. Little aid was afforded them by his many books, those records of a scientist's work most often expressed in a creative writer's language, or even with a novelist's freedom—and again inhospitable to objective analysis. It is plausible therefore to believe that William Beebe the scientist may well have suffered in consequence, his fame and his creativity working against his professional reputation; but such were the hazards of the man's chosen way, as no doubt he knew from the start, or soon learned.

There are many reasons for wishing to be known and read beyond professional circles. In a light mood Beebe once called all his books potboilers, written to finance forthcoming expeditions. But no one who reads Beebe with any attention can credit so frivolous an avowal; and the same is true of the work of Carson and Goodall and Schaller and many another, past and present, whose study is the world of living things. If a book sells widely or an article is well received, the author is bound to be pleased, but a dedicated naturalist is moved by more compelling reasons. There is news to tell of what has been sought and found; and there is also challenge and joy, beauty and wonder and devotion to be imparted in words that capture and hold them best, and perhaps endure the longest.

:11 The Finality of Words

> To sum up, I present an *ideal* equipment for a naturalist writer of literary natural history: Supreme enthusiasm, tempered with an infinite patience and a complete devotion to truth; the broadest possible education; keen eyes, ears, and nose; the finest instruments; opportunity for observation; thorough training in laboratory techniques; comprehension of known facts and theories, and the habit of giving full credit for these in the proper place; awareness of what is not known; ability to put oneself in the subject's place; interpretation and integration of observations; a sense of humor; facility in writing; an eternal sense of humbleness and wonder.
>
> *The Book of Naturalists*

Here indeed was a summing up, and one given by an authority.[1] With only two more books of his own to write, Beebe was offering in this passage the creed by which he had written all the others. When he spoke of "a naturalist writer of literary natural history" he was being precise, not redundant, and the attributes he listed are those he saw in himself. By then perhaps he knew that it would be as a writer-naturalist that he would best endure, cherished more as an adventurer than as a scientist and living longest in the books recounting his deeds. Beebe had already quoted Kipling as saying, "The magic of Literature lies in the words, and not in any man." But words as such bestow no form; what is magical resides in the way they are chosen and arranged—the style. In 1753 the grandiloquent naturalist Buffon had put it memorably when he said, "Le style est l'homme même." By his style the man himself is delineated and defined, and presented as he wishes to be known. From the start this had been William Beebe's understanding and concern.

References to stylistic matters appear frequently in Beebe's works, sometimes in mock apology for puns or disgressions or idiosyncratic usages, elsewhere in admiring references to other writers, and occasionally in direct citation of authorities, the *Oxford English Dictionary* or *Fowler's Modern English Usage* or George Crabb's *English Synonyms*. In doing book reviews in the 1920s and 30s for the *New York Times* and *The Atlantic Monthly* Beebe had dealt here and there with style as well as content; so by the time he undertook to edit *The Book of Naturalists* he could draw upon long experience in judging both his own work and that of others. His introductory passages on the authors he selected are replete with analytical comments both apt and assured; Beebe speaks not as a man venturing into alien territory but as a seasoned critic, confident of his learning and the authority bestowed by forty years of successful authorship. Thus he describes Edward Topsell's translation of Gesner as "majestic" and notes the "delightful discursive style" of Leeuwenhoek and the "slightly stilted" diction of Charles Waterton; he suggests that the paragraphs of John Muir contain "slightly too many words;" and of Gilbert White he says: "This gentle English curate was perhaps the first naturalist who clothed his observations in a real literary style."

William Beebe came far too late to claim a similar primacy in America. Alexander Wilson (a poet before he became an ornithologist) and the protean Audubon had preceded him, as well as William Bartram, the American naturalist and explorer read with interest by Wordsworth and Coleridge and praised by Thomas Carlyle for his "wondrous kind of floundering eloquence." And for Beebe there had always been his great favorite Thoreau, and in his own time John Burroughs, a boyhood idol and later Beebe's companion on an outing in Florida. Nor should one forget his friend Roosevelt, who brought literary awareness to nearly everything he wrote, not the least his outdoor essays. So Beebe had many worthy models to study among writers already established in his field.

Not all of his possible models by any means were avowed students of nature. Beebe's reading was markedly eclectic, as a glance at the literary references in his books will show. They range from Greek myths, the Bhagavad-Gita, and the Bible through Shakespeare, Swift, Byron, and Shelley to Chesterton and Wells and Conan Doyle. Among American writers are Twain, Poe, Artemus Ward, Finley Peter Dunne, and Sinclair Lewis. Except for obvious borrowings—"without form and void," "each . . . after its kind," "dusty death," "trippingly on the tongue," "a swarm of golden bees," "the heart of the night's darkness," and so forth—there is no way to judge these sources for their influence on Beebe's work. What is clear is that Beebe read other writers with an active awareness of their style, but he truly aped none of them. Any first rate writer is finally his own man, and Beebe was no exception.

One class of writers Beebe held in special esteem, the tellers of exotic tales—obviously Kipling and Henty and Conrad and Stevenson, but with a further group one might not readily suspect, the fabulists and even the writers of children's fantasies and fairy stories. These latter writers are more often cited and quoted than any other. In one passage Beebe praises not only *The Jungle Book* but *Alice Through the Looking Glass,* Milne's *The House at Pooh Corner* and Kenneth Grahame's *The Wind in the Willows,* clearly undismayed by the outright anthropomorphism in all these books. Elsewhere he speaks of one's first reading of *Alice* as a thrill rarely equaled, and in *Jungle Days* he describes the mysterious life of a mangrove tangle as fit to be imparted by "Carroll or Dunsany or Barrie or Blackwood." That is to say, there are things so fabulous in nature that they are best conveyed by such traffickers in wonder as these four.

So long as *Peter Pan* is staged or *Alice* read, two of Beebe's authors will be known and cherished; but Algernon Henry Blackwood (1869–1951) is a name familiar to few but devotees of the occult. And as for Lord Dunsany (1878–1957)—by all odds Beebe's favorite—he has gradually declined from a modest literary fame into neglect and lately virtual oblivion. It is mildly unsettling to discover that this "castle Irish" playwright and romancer was once seriously listed alongside William Butler Yeats and John Millington Synge; but a look at his work confirms the justice of his present obscurity. Momentarily at least, it also argues unhappy flaws in William Beebe's understanding; for it was not just when Dunsany was enjoying a vogue, but long afterward that Beebe held him in high regard. It should be noted, however, that Beebe's interest centered on the early tales and plays, and did not extend to Dunsany's voluminous later work.

Edward John Moreton Drax Plunkett, eighteenth Baron Dunsany, aspired to fame as both a dramatist and a spinner of strange tales in prose. Beebe could have encountered him as early as 1905, when his first collection of fiction, *The Gods of Pegana,* was published in London. Later he met Dunsany, and they became correspondents and friends. On the evidence of references he makes, Beebe particularly admired *The Book of Wonder* (1912), subtitled *A Chronicle of Little Adventures at the Edge of the World.* Its fourteen brief stories are exercises in the exotic, the occult, the supernatural; many are well enough crafted, but none has any great distinction. About this time Dunsany was also writing plays, somewhat less fanciful in their settings, but again remote from the common world. Working in either form, he persistently recounts odd doings and contrives strange names for his scenes and characters: Thangobrind, Thek, Mluna, Bombasharna, Slith, Slorg, Shard, Eznarza. The frequency of their appearance in Beebe's work shows his fondness for such literary inventions, and his enthusiastic references to Dunsany himself confirm Beebe's devotion.

To belabor Dunsany at this late date is neither fair nor useful. The attenuation and impoverishment of the Romantic spirit in the late nineteenth and early twentieth centuries was not his doing; he was a legatee but scarcely an instigator. One hears distorted echoes of Coleridge and Poe in Dunsany's description of a heroine whose "beauty was as a dream, was as a song; the one dream of a lifetime dreamed on enchanted dews, the one song sung to some city by a deathless bird blown far from his native coasts by storm in Paradise." Or: "Bells pealed in frantic towers, wise men consulted parchments, astrologers sought of the portent from the stars, the aged made subtle prophecies." Or again: "Whether it was chance that brought them through the forest unmouthed by detestable beasts, none knoweth." The reader perceives with some reluctance that these passages are not intended as parody.

The wonder is that Beebe, given his perdurable admiration for the author of such misbegotten rhetoric, should have escaped contamination. His own writing style matured in the very years when he was discovering Dunsany, but in quite another direction, toward increasing grace and clarity, surer discipline and subtlety and wit. Plainly, then, Dunsany's major appeal lay not in his style but in the escape he afforded into his "world of fancy." In *The Laughter of the Gods,* a play of 1917, the main characters flee from the crowded city to the jungle retreat of Thek; in *The Tents of the Arabs* (1914) a ruler renounces his throne for the love of a desert girl and the freedom and simplicity of a nomadic life. Many of Dunsany's tales concern a similar flight from the boredom of the actual into the excitement or enchantment of the fanciful. In a way, Beebe's life as a naturalist had followed the same pattern. Beebe was quite ambivalent toward his home city of New York, at times happy for its metropolitan verve and glitter, at other times repelled and anxious for escape. And when he fled, it was to a world which promised wonders.

That Beebe's world of nature was obtensibly one of reality and Dunsany's one of fable is not the paradox it may appear. "I think," wrote Beebe in 1922, "that some day I shall find that Dunsany's tales are sheer fact, for I have just finished studying a parasite on the gills of a fish which in its lifetime does things as strange as any in the 'Book of Wonder.' "[2] Beebe was indeed a gatherer of facts, Dunsany a conjurer of visions and dreams; but in Dunsany and Carroll and others Beebe found that "eternal sense of . . . wonder" which informed his entire scientific career. Professionally he was tied to the observable, but he was forever unwilling to keep himself emotionally or intellectually within such narrow bounds. At times he escaped from the empirical through substitute activity—on Haiti by strenuous games of tennis and a lofty highball or two, at Kartabo by staging a birthday party for himself on a date remote from July 29—or by reading aloud from Dunsany's *The Chronicles of Rodriguez* "to a circle who demanded it over and over again."[3] Another man might have taken refuge from professional tasks by writing not merely fiction (as in *Pheasant Jungles*) but fantasy, as with the mathematician Lewis Carroll. Instead Beebe allowed Dunsany and the others to provide this outlet for him, recalling both a child's delicious apprehensions and an adult's enduring sense of the strange and wonderful, the inexplicable, the immanent but undisclosed.

So these purveyors of fantasy served far more as Beebe's personal resources than as models or influences on his writing. He cherished their vision because in fact he shared it; but in his own work he was wise to keep it hidden, however powerfully it may have moved beneath the surface of the words. The substance of his work did not directly reflect such writers, nor did his style, except perhaps for a scattering of archaisms. For Beebe nature held much of the fabulous, but he did not write of it as fable.

Though Beebe's favorite authors in this special area were European, and the bulk of his literary references elsewhere were also foreign, there is much about his outlook and style of writing that is clearly American. Both early and late his idiom was the nature essay, a form so well developed in America that it seems a native growth. Apparently it is a national trait to come afresh to nature, to encounter her *de novo,* and from the meeting derive that kind of personal response which the essay embodies best. There may be good historical reasons for this. When the English cleric Gilbert White brought out his *Natural History of Selborne* in 1789, his native land had been settled for millennia, and indeed Selborne parish itself had a history that White could trace back 700 years. But in America the year 1789 marked the ratification of the constitution under which the new nation was setting forth to claim its destiny, along with the larger part of temperate North America. At Selborne there seemed little left to discover; its "natural productions and occurrences" could be observed and commented upon, but their presence and identity had long been known. By contrast, there was not a rural parish in all the United States with a history even remotely to be compared to that of Selborne, nor one with its natural life so widely and thoroughly studied. Devoted naturalist though he was, Thomas Jefferson in his *Notes of the State of Virginia* (1784) was able to list only a fraction of the birds and quadrupeds of his home state. Whether here in the oldest of the former colonies or in the vast areas yet to be explored and occupied, the rest were still to be discovered.

This was to be the paleface American experience with the natural environment of the continent: exploration and discovery, settlement and exploitation, and not infrequently outright pillage. Amid this huge enterprise the naturalists were a tiny group with a different aspiration, to seek and find, to observe, describe, acclaim; and from the earliest times, to cherish and protect. Neither law nor custom effectively supported them. They had no power but that of words and pictures, nothing to offer but new knowledge and new responses. So it was with Mark Catesby and William Bartram, Alexander Wilson, John James Audubon, George Catlin, Clarence King, John Wesley Powell, John Muir—and as one of the last, Theodore Roosevelt, even as a youth both an inveterate hunter and a budding naturalist. When Hector St. John Crèvecoeur asked his famous question, "What then is the American, this new man?" part of the answer was woodsman, plainsman, explorer. It was scarcely by accident that our first considerable novelist should have turned to the American wilderness as the setting for his best work, the five novels that make up the *Leatherstocking Tales,* nor that he should have lamented its passing before the westward tide of settlers. There is no more wonderful novel in our literature than

Huckleberry Finn, and here again a wilderness magic pervades the tale, the power of its spell drawn from the majestic ancient river. Thoreau said that "in Wildness is the preservation of the world"[4] —an exceedingly American statement, springing as it does from a national experience that has touched us all.

Such was Beebe's heritage, but he came late in time to his legacy. He could not share with James Fenimore Cooper the trackless forests around Glimmerglass, for they were long gone, nor could he hunt with Audubon amidst "the grandeur and beauty" of the Mississippi's "almost uninhabited shores" or sail with Alexander Wilson down the Ohio of old. Not even the primeval west of John Muir was still to be explored; at about the time William Beebe was attaining his majority, Frederick Jackson Turner was proclaiming the end of the American frontier and reflecting upon its potent influence in shaping our national character. Whether Cooper's picture of the wilderness was authentic or Turner's thesis valid is beside the point; the pervasiveness of the experience, real or romanticized, is beyond dispute.

So it was not merely by chance that William Beebe should have spent most of his life in search of wilderness. If twentieth-century America denied him the wild places he sought, the strenuous challenges, the new and fascinating forms of life, then he would go elsewhere. Thoreau could refresh his spirit by excursions to Cape Cod or the "grim and wild" Maine woods; Beebe, much later, sought the jungles and enchanted islands and ocean depths and the distant reaches of the East. Ostensibly the purposes of the two men were different, the one philosophical, the other scientific; yet both reacted with that kind of autobiographical immediacy which has always marked the American literary response to wild nature, and is best conveyed in the relatively flexible form of the essay.

Had Theodore Roosevelt lived ten years longer, he might have become Beebe's most important literary critic. As it was, he provided an introduction to *Tropical Wild Life* and a foreword to the second printing of *Jungle Peace*. He demonstrated in these two pieces how well he understood Beebe's growth as a writer. The first classified Beebe's work with that of Audubon and "the best scientific books" of Darwin and Waterton and other English naturalists; the second not only drew a clear line between *Jungle Peace* and Beebe's earlier work, but declared this book to be "a new type of higher literature," superior to Charles Waterton's tales of exploration and comparable in its charm of language to the essays of

Robert Louis Stevenson. In short, Roosevelt recognized that Beebe had entered a new and more creative phase of his work, and he would no doubt have agreed with George Reuben Potter when that critic stated in 1929 that "[Beebe's] significance to literature is unique in that he is the one professional zoologist of the first rank whose writings hold a high place in pure literature."[5]

By this time Beebe's writings amounted to about two-thirds of all the works he would produce, and afforded a valid basis for such literary judgments. And it is clear that the four jungle books already published offered the most persuasive reasons for placing Beebe on such a high plane of accomplishment. They were the most carefully wrought, the most consciously literary of his works, whether composed of vivid field reports, as in *Edge of the Jungle* or *Jungle Days,* or of the fictionalized adventures which make up *Pheasant Jungles* and much of *Jungle Peace.* These are by conventional standards his best creations, and nothing he was subsequently to write would surpass them or radically alter the judgments already made. In a few years *Nonsuch: Land of Water* was added to the list of comparable works, and later Beebe's last four books of essays, beginning with *Zaca Venture* and ending with *Unseen Life of New York.*

But in the Beebe canon there are other books with different attributes, calling for judgments to be made by other criteria. It is true, for example, that Beebe's two volumes of Galapagos adventures and his major undersea book, *Half Mile Down,* do not excel in the fashion of those mentioned above. At times diversity of subject matter makes for diffuseness, and multiple authorship threatens unity of style. There is a tendency to sprawl, suggesting the need for a stronger editorial hand; in places there is a haphazard quality to the writing, indicating haste and even indifference to literary effects. This was the most crowded creative period of Beebe's life, with fifteen volumes appearing between 1919 and 1934; hence these books may be excused from serious appraisal as having been written too swiftly to have been written well. The Galapagos books, among the longest of Beebe's works, appeared only two years apart, with *Jungle Days* and the two-volume *Pheasants: Their Lives and Homes* sandwiched between. The contrast with other books recounting sea voyages is marked; *Zaca Venture* and *Book of Bays,* considerably shorter and more integrated works, came out in 1938 and 1942, with no major publications in the interim.

Yet if they do not fit comfortably into categories nor always satisfy customary standards of taste, *Galapagos: World's End* and *The Arcturus Adventure* are far from worthless productions. In them Beebe tells of his search for wilderness far beyond the borderlands of the Caribbean, in the reaches of the equatorial Pacific; and in them also may be found exploits and discoveries and wonders which the other sea books cannot boast. These Galapagos books swell out of proportion partly because they are so copious, so prodigal of content. From the riches of the two voyages to the Encantadas, Beebe and his associates made haste to include the most exciting times and display the most vivid scenes and pictures; and if the resulting fat volumes are somewhat ungainly in form, they are bountiful in yield and winningly zestful in spirit.

Beebe's two popular works on Bermuda offer contrasts even more instructive and significant. As he had in his early jungle books, in *Nonsuch: Land of Water* Beebe again was writing for the magazine market, and doing it exceedingly well. These essays are more than technically expert, they are intellectually challenging. One is happy to find Beebe returning to the stylistic sensitivity of, say, *Jungle Days,* and adding to it new clarity and depth and subtlety of understanding. And one cannot help admiring the range of knowledge informing these essays, the reward of thirty years of work in natural science, here offered with supple grace and a sophisticated awareness of the diverse life of this one small island.

The other Bermuda book for the popular audience was *Half Mile Down*—and the setting is about all these two works have in common. Once again we are back to relatively hasty writing and scattered organization, demonstrated by the fact that much of this volume is written by others besides Beebe, a good deal concerns history rather than present time, and some of the undersea work described dates back several years to Haiti or the Galapagos. Although the historical sections are enlivened by touches of the author's playful wit and engaging diction, they remain fairly straightforward. The voluminous appendices—a good third of the book—are similarly shy of creative spirit. The heart of the volume, concerning the bathysphere dives, is mostly given over to narrative, much of it relatively unadorned. These chapters are quite diverse in length and cover three separate periods, the descents made in 1930, 1932, and 1934. One is led to suspect that *Half Mile Down* was not so much made as just allowed to grow.

And yet when *Nonsuch* and *Half Mile Down* are judged side by side it is the former that is diminished by the comparison. The literary and even intellectual virtues of *Nonsuch* have not the ' to outweigh the wonders of those bathysphere descents, howeveı plainly or even gracelessly told. Behind the narratives of the dives, the fixed pattern of descent and return, lie mysteries to be fathomed and inescapable dangers to be confronted. Though his vessel was a steel ball on a steel string and not a swift and black-prowed ship, Beebe's voyage to the unknown deeps was in its way Odyssean. One wishes that the telling could be called Homeric. From thirty-odd ventures into the black wilderness came a few tales only, gathered to form not an epic of discovery but mere parts of a larger book. If by their magic the book is elevated and sustained, it magical qualities arise from the deeds themselves and not the enchantment of words.

There are reasons for this which go beyond carelessness or haste. Speaking of the deepest of all his descents, Beebe said: "Adequate presentation of what I saw on this dive is one of the most difficult things I have ever attempted. . . . Only the five of us who have gone down even to 1000 feet in the bathysphere know how hard it is to find words to translate this alien world."[6] *Alien*—here is the key word. The sea at a thousand feet or deeper is unimaginably strange, and in all his life Beebe spent only fleeting hours there instead of days or years; he had not loved and rejoiced in it for half a century, as in the upper world of sunlight and birds and jungle green, and would never do so. Alien it was and alien it remained: forever and intensely dark and cold, airless, odorless, nearly soundless, seemingly changeless—such was the nether world to which Beebe briefly journeyed, as mythic heroes to Hades. But here were no spirits of the departed, no damnations or evil spells, no demons or imps, no Grendel or Cerberus or Charon; nor was this in the least like Alice's world beyond the looking glass or Dunsany's realm of fancy. Indeed it was not fanciful at all, but natural. No man had ever seen what Beebe and Barton first saw; in a strange way, however, it was all comprehensible and scientifically explicable. Whatever Beebe saw clearly he could identify, at least as to phylum, and even his most impressive new species, the six-foot Untouchable Bathysphere Fish, he could confidently list under the phylum Chordata, subphylum Vertebrata, class Pisces, subclass Toleostei, order Isospondyli, and doubtless the family Melanostomiatidae.

Still this great fish *was* untouchable, and so was every other deep sea creature discovered through the six-inch windows of the fifty-four inch bathysphere. Within so narrow a compass, to see anything at all was a matter not of skill but of accident—a sudden, silent appearance, a brief moment of observation, then a vanishing, in all too many cases forever. Here indeed were natural mysteries in potentially unlimited numbers, but tantalizing, frustrating, and at last nearly barren, whether for science or art. On land or even in the coral shallows, a creature can be seen and perhaps heard as it approaches, as it passes near or comes to rest, as it departs. It may return or be pursued and collected; it can be studied in relation to its fellow creatures and followed through successive stages of its life. All of this and much more is available to an assiduous observer—and none of it to Beebe in his bathysphere. Far more than Garner in his cage, Beebe was a prisoner within a shell of steel, deprived of the use of all but sight, and that one sense narrowly confined to a small aperture rigidly fixed and a beam of light even more delimited.

Finally there was the aspect that might be called technological, particularly unsettling to one of Beebe's special temperament. The bathysphere for all its utility and efficiency was something of a mechanical contraption, scarcely congenial for a man who admitted his imperfect sympathy with machines. Inside it was cramped and everywhere concave, inevitably causing muscles to ache and feet and legs to go to sleep. Furthermore, it was cluttered with equipment (some of which might shift in rough weather) and far from silent; there was the sound of the air-circulating system, and even worse, perhaps, the constant need to talk to those on deck and receive questions in return. Beebe and Gloria Hollister had agreed that no more than five seconds should pass between comments; anything greater would be taken as a danger signal. Although Beebe had discovered very early that voice communication was essential for morale, it was also intrusive and unremitting. Outside was the endless dark of the limitless deeps, but within the bathysphere little repose.

In "Jungle Night," the final chapter of *Jungle Peace,* Beebe wrote of the wonders of solitude and quiet and exquisite sense perception which he had enjoyed in nocturnal wanderings through the moonlit forest near Kalacoon. If there is nothing comparable in *Half Mile Down* it is because such responses were denied him by the very nature of the bathysphere itself, leaving the wonders to be inferred from the forthright account he gives of these unique adventures. Though we may regret that Beebe chose not to recount the times of the bathysphere in the style of *Nonsuch* or *Jungle Days,* we must accord him the right to make that choice, and recognize its creative validity.

Jungle Peace was more than William Beebe's first book of high literary merit; it was also, as noted earlier, his first tentative venture into fiction. He tried to impose a loose chronology based on a quest for peace after battlefront perils had gravely unsettled him, and he used his war experiences as a recurrent theme affording unity to essays otherwise diverse. Neither effort was particularly successful, although early in the book both showed a certain promise. In any case *Jungle Peace* was Beebe's initial attempt to use such themes creatively, not in straightforward narrative or descriptive prose, but obliquely, by selective emphasis and implication—in a word, fictionally.

Always excepting Thoreau, the most profound use of nature in American prose literature has been fictional. We have no "nature novel" to compare with W. H. Hudson's *Green Mansions,* but two of our greatest works, *Huckleberry Finn* and *Moby Dick,* are deeply engaged with the natural world both as the setting for major action and as a source of pervasive meaning and inescapable power. Melville's Captain Ahab, unnaturally hating the brute creature that had maimed him, challenged the natural order and came to disaster, drowning in the deep with the dying whale. Huck and Jim fled from bondage down the great river, finding joy and peace on its tide, even as it bore them irresistibly deeper into the stronghold of slavery where no escape was possible. Later novelists—Hamlin Garland, Frank Norris, Jack London, perhaps Theodore Dreiser—found in Darwin and Spencer and Nietzsche the basis for a kind of biological determinism which gave the natural world even more encompassing control. It is suggestive that their fiction is called "naturalistic."

Much of this outlook reappears in the work of John Steinbeck, who was trained in marine biology at Stanford—where no doubt he first encountered Beebe's work—and who chanced to follow much of the route of Beebe's first *Zaca* voyage when he made his own collecting trip in the Gulf of California in 1940. From this journey came the substantial volume *Sea of Cortez* (1941)[7] and *The Log from the Sea of Cortez* (1951),[8] a narrative abridgment. There is more to *The Log* than narrative, however. At times Steinbeck is the coolly scientific collector; elsewhere he is the natural philosopher, unwilling to view the world as something bestowed to serve the ends or enhance the self-esteem of man. He insistently rejects "our iron teleologies" and "the leading-strings of a Sunday-school deity" for an ecological pantheism affirming "that all things are one thing and that one thing is all things—plankton, a shimmering phosphorescence on the sea and the spinning planets and an expanding universe, all bound together by the elastic string of time."[9] Among the half dozen greatest figures to comprehend and embrace this view Steinbeck includes Albert Einstein and Charles Darwin.

In the thirteen stories collected in *The Long Valley* (1938) Steinbeck had dealt with nature less mystically, more concretely, but with no less awareness of her inexorable force. Perhaps the harshest of the stories is "Flight," in which young Pepé Torres (whose father had died years before of a rattlesnake bite) kills a man and is hounded to his death in the mountains. Pepé is cut down by a rifle bullet, but he has already lost his struggle with natural forces, the pitiless slopes that offer no water or food to sustain him, no relief from the sun, no cure for the infection which disables his right arm. Once as he crawls for cover Pepé is nearly bitten by a rattlesnake, and near the end a mountain lion sits and quietly watches him, moving silently away only at the sound of distant pursuers. As there is no malice in these threatening wild creatures, there is also no mercy; these are human attributes, not so much denied by nature as absent from her order of things. This view of the natural world may be found in many other places in Steinbeck's fiction.

Nature's sexuality is something John Steinbeck heartily appreciates and in one passage in *The Log* celebrates through the sense of smell, unblushingly comparing the rutting of goats and of men. In the work of Ernest Hemingway sex and nature are linked in a more complex way, and even more insistently. His important early experiences with both came in the woods of Michigan, the natural setting for three stories of sexual awakening, "Ten Indians," "Fathers and Sons," and "Up in Michigan," and two concerning the aftermath of early love, "The End of Something" and "The Three-Day Blow." As with the Beebes in Mexico, in these stories outdoor nature and youthful desire are often joined—but with Hemingway there is more. Natural simplicity is shadowed here by psychological complexity, the latter often dominating the former, despite the vividly imparted charms of the setting.

Again in "Big Two-Hearted River," a long fishing story also set in the Michigan back country, Hemingway adds a further aspect to experiences common to both writers. To soothe the pain of lost love and the psychic wounds of war, Beebe had sought and found remission in the tropical jungle; Nick Adams, similarly burdened, returns to his cherished river but cannot find a similar release. The Hemingway scholar Richard Hovey perceives the reason when he finds that Adams in fact resists nature's comforting embrace, the solace of her remembered joys. Though the outward scene is described in almost naturalistic detail, Adams "dare not look within"; and when at last he does, instead of deliverance he finds a baleful symbolism.[10] He has fled to nature but he has not escaped; the terrors of memory, whether martial or sexual, have been reinforced as well as appeased by a return to the wild. To Melville the whiteness of the whale was above all else appalling: "Though in many of its aspects this visible world seems formed in love, the invisible spheres were formed in fright." So too for Ernest Hemingway: what was visible in nature was by no means all that was there.

It is this aspect of the hidden, the brooding, the ambiguous, the threatening, and above all the inescapably primitive in nature that fiction, and perhaps American fiction in particular, can impart best. William Beebe had written book after book of essays devoted enthusiastically to nature's "visible world . . . formed in love"; in *Jungle Peace* he had gropingly sought "the invisible spheres," and in *Pheasant Jungles* he ventured as far as he would ever go within their realm. In so doing he moved beyond his familiar and widely accepted role as America's foremost writer of nature adventures, and sought to join those who had written and were still to write of nature in quite a different way, more freely and creatively, and more profoundly also.

In Beebe's introduction to *Pheasants: Their Lives and Homes* there is this bald declarative sentence: "In fact, nine men lost their lives in the course of this expedition." No other information is given here, and no such statement appears in the book which presumably should deal with these deaths, *Pheasant Jungles.* In that work only one of the nine can be identified without question —the "Chinaboy" servant Lanoo, who died (allegedly of superstitious dread) after Beebe had lanced a felon on his finger. The "renegade" Beebe tells of shooting was in no way part of the expedition, but perhaps can be included in the total. Months earlier, in Egypt, desert robbers had attacked the Beebe party, apparently leaving behind one of their band dead or wounded. The "strange tribesman" whose half-eaten body was found near Sin-Ma-How was a mere visitor; and then there were the defective Kachin children driven from their homes to die in the mountains near the camp, but Beebe gives only sketchy details and no figures, and he has nothing to do with their supposed deaths. Drojak's eight victims are wholly remote from expedition affairs, except as severely truncate waistband ornaments. *Pheasant Jungles,* then, is a book shadowed by half a dozen deaths which are unavowed and unexplained. Whether fictional or real, these deaths impart to this book a somber or even tragic quality found nowhere else in all of Beebe's accounts of his wanderings.

Although in his earlier report on the pheasant expedition Beebe only hinted at such things (and of course allowed Mary her proper place as a full participant), two profoundly distressful experiences were presently to work great changes in his spirit, and in his writing as well. First came Mary's despairing flight and suit for divorce, immediately causing her banishment from the text of the *Monograph,* and soon thereafter from the magazine articles in which the pheasant journey was reformulated, especially with the introduction of more difficulty and danger, whether from natives or from wild creatures and threatening natural forces. Then in the midst of this retelling came the Great War, bringing to Beebe a far more intimate acquaintance with death and the risk of dying than he had ever known.

Amid the somber reports Beebe made of his wartime experiences, two in particular stand out. "Animal Life at the Front," already discussed, offers bizarre contrasts between human death-dealing and natural adaptation to such terrible circumstances. Not merely house sparrows and rooks and skylarks manage to carry on their lives; so also do the battlefront scavengers, the rats and the flies whose maggots feed on the unburied dead; and wolves have returned to the forests near the battle lines, driven by hunger. The second article, a little-known piece called "A Red Indian Day,"[11] also mentions the wolves, but involves Beebe even more directly in the strange environment of the forward trenches. The Indians of the title are Algonkins and Iroquois—one smilingly asks Beebe, "Looks as if we'd climbed out of Cooper, doesn't it?"—who go on a night scouting raid, taking Beebe part way, and return with a captured German. Back in the dugout they undergo an artillery barrage that causes the water covering the floor to quiver.

This same article contains a passage which for both discipline and intensity is Beebe's most compelling statement on the war. After spending a troubled night in a French farmhouse back of the lines, Beebe had walked in the early dawn to a low hill nearby to watch and listen, anticipating "something big, something tremendous, which was about to happen":

> Then came to the senses an anticlimax, but to the imagination an awakening, staggering in its contrast and significance. Two sounds broke the stillness, penetrating the fog with difficulty, each subdued, mist-muffled. First a faint, low *kr-krump*. A month later, and it might have been a small frog in a distant pool. Then a soft, liquid gurgle, a virile contralto note—the first vocal call of life to the dawn; a contented hooded crow talking to himself from his roost among the maze of beech-twigs. The setting seemed perfect: the dawn of a late winter day in France, the voice of the bird and the frog, and the mist-borne fragrance of the freshly turned mould. How happily Burroughs or Sharp could phrase it; how the great god Pan would enjoy it! At last I could keep up the deceit no longer. Let Pan keep his bird and the smell of the mould—that was his: but the low hollow note, the faint *kr-krump*, so brief that it seemed almost a trick of the ears, this was a velvet glove of sound made by no batrachian, and Mars in his most hideous accoutrements strode forth in my imagination and claimed it. At times I have been in a perfect hell of artillery fire, and I know of no instance where the sound of a gun affected me as did this single distant *kr-krump*, miles and miles to the north. Here it came gently, a soft antiphony to the content note of the crow; there it was a volcano of flying steel splinters and hideous fumes, tearing through flesh and wood and mud, destroying all life, human, animal, and plant, demolishing even the inorganic face of nature, the dissipating smoke revealing a landscape comparable only to that on the surface of a dead moon.[12]

Thus Beebe starkly contrasted nature as he had known it with nature atrociously devastated by war. The kind of benignity offered by John Burroughs or Dallas Lore Sharp could not encompass the new reality—this much Beebe understood. But he still had one more big step to take. Nature in this passage is the tortured victim of deathly humanity, and Beebe is only the anguished witness. The sense that nature itself could be malign, and could threaten not only human life but the human spirit as well, he had yet to entertain. Warfare, awesome and terrible, took him part of the way; it added a tragic dimension to his understanding, and no doubt prompted him to recall and evaluate anew (or even to invent) certain experiences of his long and troubled expedition to the East.

And so we have in *Pheasant Jungles* those instances of dread and fearful alienation from nature found almost nowhere else in William Beebe's work. The first of these is low-keyed and briefly told, but not to be ignored. The expedition has reached Myitkyina, and Beebe, looking at nightfall at the wilderness across the Irrawaddy where he will soon go exploring, is seized by "a fit of pessimistic terror" at the prospect, realizing he has "so little knowledge of its dangers, whether of animals or of men. . . . I always go through this stage on the eve of any new undertaking, and while it lasts it is very real and very terrible."[13] A good night's sleep is enough to dispel his fear and permit him to go on; but when he reaches Pungatong a week or so later, his disability returns in a far more crippling form, and his recovery takes much longer.

"The thought of going on was impossible. I hated pheasants, the jungle and all its inmates . . . my only desire was to run back to Rangoon, to America, to my home as fast as possible and never think pheasant again. . . . I lit my candle lantern and roamed about the bungalow trying to estimate in how few days I could make my return."[14] It was then that Beebe chanced upon a store of cheap novels and began compulsively to devour them. It took three days of reading, sleeping, and idling to restore his spirits to the point that he was willing to proceed toward the wild mountains of the Burmese borderland, where he would undergo the most threatening physical attack of his entire eastern journey.

The last onset he describes was perhaps the most terrible. Beebe relates how some wandering natives encountered his expedition one moonless night along the Pahang in Malaya and asked to be hired, and how he called out in the darkness that three of them should be taken on. He had barely stepped ashore the next morning when the three confronted him, and he saw with profound shock that all were lepers, hideously ravaged by their affliction. This horrendous experience seemed to bring on a fever, making all the more oppressive the humid silence of the jungle as Beebe sat and listened for the notes of birds he was seeking. Instead it was a host of leeches he heard, or seemed to hear, advancing upon him from all sides with a slithering sound that at last he could not bear. He stood up and shot twice into the foliage overhead, and for a time his fear left him. But then as he waited tensely for a group of peafowl to come within range, he felt something strike his back and he leaped screaming for a tree and climbed ten feet up in utter panic.

"A tiny squirrel—one of the little dwarfs scarcely as long as one's hand—had jumped on my back, and I had reacted as from the charge of a buffalo."[15]

This seemed to clear his head and quiet his fever, but there was still more to come. Having stalked and brought down one of the birds, Beebe marked the direction of its fall and walked directly toward it, only to step over the edge of a concealed ravine and end up suspended many feet from the bottom in a tangle of needle-sharp rotan thorns. It took him an excruciating quarter of an hour to descend through this merciless net of pain; then at last he reached the ground and recovered his gun, found his bird, and made his way wearily to camp.

In the recounting of these deeply unsettling experiences in *Pheasant Jungles* William Beebe manifested as in no other book an awareness of nature's dark side, those "invisible spheres" known so well to Herman Melville. Whether these were actual events or fictional creations is of little consequence. They appeared as he chose they should, and in the form suited to his purposes. They reveal to the student and biographer of this extraordinary man depths and ranges not discovered before, and suggestions of things yet to come. Beebe at fifty apparently had found a new voice and set a new goal—one which might lead him beyond his naturalist's calling into the writing of fiction, for which he had now demonstrated a genuine talent.

That he did not choose such a course we know from his subsequent books. The dark intimations of *Pheasant Jungles* were never to be fulfilled. What Beebe had learned and imparted in this book he would not state again in comparable terms. The Great War was behind him; if one marriage had ended, another was beginning; and the eastern journey was receding into the realm of cherished memory. Beebe understood that this was true and in his own words confirmed it. "A Red Indian Day" spoke of a battlefield so frightfully desolated as to look like the moon's dead face—but the next sentence, ending the story, is this: "The crow called again, clearer this time, less liquid, as if the thinning mist had dried from his throat; and quickly the fog drifted off toward the valley, and now the tinkle of the brook became audible again, and I walked rapidly down to the thatched farm-house." And here is the passage immediate following Beebe's trauma with lepers and leeches and thorns: "The following day I had to rest; but when again I made my way through this same jungle, I saw it only as a place of wonder, of keen delight, and of deepest interest."

In these two sentences Beebe acknowledged something exceedingly important about himself. Whatever the pain and threat of such evil events, whatever the insight provided and the doubts aroused, the outcome was not at issue. The years simply would not be denied. As a boy in New Jersey he had felt a "very real terror" at the vastness and loneliness of nature, and still he had joyfully pursued butterflies and birds in the parks and fields and the Orange Mountain woods, and had discovered a career which led him to the distant places of his boyish dreams. The story of *Pheasant Jungles* is the story of the greatest adventure of them all; to deny that adventure would have been to reject or deprecate the career he had led and the man he had become. William Beebe had passed through trials of body and spirit which he did not choose to disavow; in the dim forest he had sensed the tiger, and knew it to be fearful. But he was a yea-sayer, a lover, a child of light, and would not see the dark prevail.

Early in 1926 the *New York Times* asked a number of famous people what they found to be the most admirable American trait, and William Beebe answered, "Undoubtedly, an enthusiasm for living to the limit." He did not speak as an idle bystander. For a quarter of a century he had been an irrepressible enthusiast for the living world, learning it, venturing forth and testing it, rejoicing in its abounding life and writing of his joy, as he would for thirty-odd years to come.

Near the end of *Walden* Thoreau speaks of a successful life as one built on dreams, a life imagined and then realized. What Beebe finally tells us in his writing is how faithful he was to his dreams, how confidently he pursued them; and this above all else was his success, a life chosen and achieved, and lived, as he recommended, to the limit.

NOTES

1. FATHER OF THE MAN

1. William Beebe, *Pheasant Jungles* (New York, G. P. Putnam's Sons, 1927), p. 97.

2. By the census of 1880, East Orange had 8,349 people, and all of Essex County only 189,819. Brooklyn that same year had a population of 566,689. Available directories for both places indicate that the Beebe family moved to East Orange in 1884 or 1885.

3. William Beebe, *Nonsuch: Land of Water* (New York, Brewer, Warren & Putnam, 1932), p. 156.

4. William Beebe, *Beneath Tropic Seas* (New York, G. P. Putnam's Sons, 1928), pp. 186–187.

5. Lee S. Crandall, "Charles William Beebe," *The Auk*, January 1964, p. 38.

6. See Clarence Beebe, *A Monograph of the Descent of the Family BEEBE from . . . John, of Broughton, England, 1650*. New York, 1904.

7. William Beebe, *Half Mile Down* (New York, Harcourt, Brace and Co., 1934), p. 8.

8. William Beebe, with G. Inness Hartley and Paul G. Howes, *Tropical Wild Life in British Guiana* (New York, New York Zoological Society, 1917), p. 160.

9. William Beebe, *The Arcturus Adventure* (New York, G. P. Putnam's Sons, 1926), pp. 340–341.

10. Crandall, "Beebe," p. 38.

11. The doctoral title used by Beebe came from honorary degrees awarded him in 1928 by Tufts College (Doctor of Letters) and Colgate University (Doctor of Science).

12. See Crandall, "Beebe," p. 39; Fairfield Osborn, "My Most Unforgettable Character," *Reader's Digest*, July 1968, p. 128.

13. William Beebe, *The Bird, Its Form and Function*. New York, Henry Holt and Co., 1906 (hereafter cited as *The Bird*).

14. William Beebe, *The Log of the Sun*. New York, Henry Holt and Co., 1906.

15. William Beebe, *Two Bird-Lovers in Mexico*. Boston and New York, Houghton Mifflin and Co., 1905 (hereafter cited as *Two Bird-Lovers*).

2. JOURNEYS WITH MARY

1. Blair Niles, ed., *Journeys in Time* (New York, Coward-McCann, Inc., 1946), p. 2.

2. Ibid., p. 3. For other autobiographical material (more or less fictionalized) see Niles, *The James* (New York, Farrar & Rinehart, 1939), pp. 9–26.

3. *Two Bird-Lovers*, p. 13.

4. Mary Blair Beebe and C. William Beebe, *Our Search for a Wilderness*. New York, Henry Holt and Co., 1910.

5. William Beebe, "The Zoological Society's Pheasant Expedition," *Zoological Society Bulletin* (New York, July 1912), pp. 763, 771.

6. William Beebe, *A Monograph of the Pheasants*. 4 vols. London, H. F. Witherby and Co., 1918, 1921, 1922 (hereafter cited as *Monograph*).

7. Details and quotations from the Beebe divorce action are taken from "copies of the Petition, Answer, and copy of what had been filed in the case" obtained from the Washoe County Court House, Reno, Nevada.

8. William Beebe, *Galapagos: World's End* (New York, G. P. Putnam's Sons, 1924), pp. 293–294.

3. THE GREAT WAR

1. William Beebe, *Jungle Peace*, New York, Henry Holt and Co., 1918; and *High Jungle*, New York, Duell, Sloan and Pearce, 1949.

2. *The Letters of Theodore Roosevelt*, Selected and Edited by Elting E. Morison (Cambridge, Mass., Harvard University Press, 1954), vol. 8, p. 1302. The letter quoted was written from Oyster Bay to Roosevelt's son Quentin, then overseas.

3. Ibid., p. 1382.

4. *The Arcturus Adventure*, p. 87.

5. *Zoological Society Bulletin* (New York, November-December 1930), p. 248.

6. See William Beebe, *Pheasants: Their Lives and Homes* (2 vols., New York, Doubleday, Page and Co., 1926), I, p. xxv.

7. See under "Items of Interest," *Zoological Society Bulletin* (New York, January 1918), p. 1577.

8. The statement by a reviewer of *Tropical Wild Life* in the British ornithological journal *The Ibis* for April 1918 that "Captain Beebe . . . now holds a commission in the Aviation Service of the American Army" is not verified by official sources. The National Personnel Records Center, St. Louis, Missouri, has no record of World War I Army service performed

by Beebe, and the National Archives and Records Service, Washington, D.C. has no record of a pension or an insurance file in his name.

9. *Jungle Peace*, p. 3.

10. William Beebe, *Edge of the Jungle*. New York, Henry Holt and Co., 1921.

11. William Beebe, *Book of Bays* (New York, Harcourt, Brace and Co., 1942), p. 226.

4. MAGNA OPERA: THE PHEASANT BOOKS

1. *Nineteenth Annual Report of the New York Zoological Society* (New York, January 1915), p. 53.

2. Daniel Giraud Elliot, *A Monograph of the Phasianidae* or *Family of the Pheasants*. 2 vols., New York, Published by the Author, 1872.

3. Lee S. Crandall, "Charles William Beebe," *The Auk*, January, 1964, pp. 37–38.

4. Paul Russell Cutright, *Theodore Roosevelt the Naturalist* (New York, Harper and Bros., 1956), p. 269.

5. *Pheasant Jungles*, p. 196.

6. Ibid., p. 114.

7. Ibid., p. 159.

8. Ibid., p. 113.

9. The original version of Beebe's first chapter in the *Atlantic Monthly* for March 1916 made no mention of tailspin or wing slips.

10. There is an odd ambiguity regarding this tribesman. Beebe's photograph opposite p. 121 shows him standing beside Aladdin at the Beebe camp, near which he was killed some days later. But the background shows a lush and level tropical grove of bananas, in startling contrast to all other photos of this mountain camp, which was surrounded by hills and minor peaks and was at night bitterly cold.

11. *Pheasant Jungles*, p. 233.

12. *Monograph*, IV, p. 194.

13. Ibid., II, p. 84.

14. Yet another version, mentioning "poisoned arrows" but indicating no retaliation by Beebe, appears in the *Monograph*, I, p. xx, and is repeated verbatim in *Pheasants: Their Lives and Homes*, I, p. xxvii.

5. THE TROPICAL JUNGLE

1. For a good succinct account of this expedition, see Paul Russell Cutright, *Theodore Roosevelt the Naturalist* (New York, Harper & Bros., 1956), pp. 241–255.

2. William Beebe, "A Yard of Jungle," *The Atlantic Monthly*, January 1916, p. 42.

3. Perhaps the most obvious interpolations are those in the third chapter, "Islands," comparing events Beebe had experienced in early 1918 on his voyage home from the war zone, with events of his "current" voyage to South America—although Beebe took no such southerly trip (because

of wartime conditions) between his return from Europe and the time the book was published, later that same year.

4. Many interesting details of the early days at Kalacoon and Kartabo are given by Paul Griswold Howes in his firsthand account, *Photographer in the Rain-Forests* (Chicago, Adams Press, 1970), pp. 3–76.

5. William Beebe, *Jungle Days*. New York, G. P. Putnam's Sons, 1925.

6. THE GALAPAGOS

1. Charles Darwin, *Journal of Researches . . . During the Voyage of H.M.S. "Beagle" Round the World* (rev. ed., 1845; reprint ed., London, Ward, Lock and Co., 1890), pp. 270–272; and Herman Melville, *The Encantadas* (1854), in *Shorter Novels of Herman Melville* (New York, Liveright Publishing Corporation, 1942), pp. 159–161.

2. Darwin, *Journal*, pp. 276–277, 289.

3. Beebe follows Robert Ridgway's "Birds of the Galapagos Archipelago" (1897), which divided the mockingbirds into eleven species. Nine forms of *Nesomimus trifasciatus* are given for these birds in Ernst Mayr and James C. Greenway, Jr., *Check-List of Birds of the World* (Cambridge, Mass., Museum of Comparative Zoology, 1960), vol. 9, pp. 447–448.

4. These related insectivorous forms (some widely distributed, some rare) include three species of *Camarhynchus,* two of *Cactospiza,* and *Certhidea olivacea.* See Robert I. Bowman, "Evolutionary Patterns in Darwin's Finches," *Galápagos Islands: A Unique Area for Scientific Investigations, Occasional Papers of the California Academy of Sciences* (San Francisco, Published by the Academy, 1963), pp. 112–113, 121. The most recent taxonomic arrangement for Darwin's finches (thirty-three forms in all) is given by Raymond A. Paynter, Jr., and Robert W. Storer, *Check-List of Birds of the World* (Cambridge, Mass., Museum of Comparative Zoology, 1970), vol. 13, pp. vii, 160–168.

5. *Galapagos: World's End,* p. 158. See also pp. 69–74, 78, 96–97, 100–101, 157, 266.

6. See Thor Heyerdahl, "Archaeology in the Galápagos Islands," *Galápagos Islands: A Unique Area,* pp. 45–51 (see note 4). The monograph *Archaeological Evidence of pre-Spanish Visits to the Galápagos Islands* by Heyerdahl and Arne Skjölsvold (Salt Lake City, Society for American Archaeology, 1956) affords a more exhaustive treatment of the subject.

7. See papers by Victor van Straelen, Misael Acosta-Solis, and Jean Dorst, *Galápagos Islands: A Unique Area,* pp. 5–9, 141–146, 147–154 (see note 4).

8. Charles Haskins Townsend, "The Galápagos Tortoises in their Relation to the Whaling Industry: A Study of Old Logbooks," *Zoologica: Scientific Contributions of the New York Zoological Society,* vol. 4, no. 3 (1925), pp. 55–135.

9. See *Galápagos: World's End,* pp. 223–224.

10. *The Arcturus Adventure,* p. 177.

11. Current Ecuadorian names for these islands are: James, San Salvador; Jervis, Rábida; Albemarle, Isabela; Duncan, Pinzón; Indefatigable, Santa Cruz; and South Seymour, Baltra. Others include Abington, Pinta;

Barrington, Santa Fe; Bindloe, Marchena; Charles, Santa Maria; Chatham, San Cristóbal; Hood, Española; Narborough, Fernandina; and Tower, Genovesa. Collectively they form the "Archipiélago de Colón."

12. David Lack, *Darwin's Finches* (Cambridge University Press, 1947), pp. 108, 110.

7. BENEATH THE SEA

1. The quote from the log is in *The Arcturus Adventure,* p. 398.

2. Succinct descriptions of dangerous marine organisms appear in Morris H. Baslow, *Marine Pharmacology* (Baltimore, William & Wilkins, 1969). The definitive work is B. W. Halstead, *Poisonous and Venomous Marine Animals of the World* (Washington, D.C., U.S. Government Printing Office, 1965, 1968).

3. While precise death rates are not easy to establish for parasitic disease, according to Peter Jordan and Gerald Webbe, *Human Schistosomiasis* (Springfield, Ill., Charles C Thomas, 1969), some 200 million of the world's population are affected.

4. *Half Mile Down,* p. 20.

5. William Beebe and John Tee-Van, "The Fishes of Port-au-Prince Bay, Haiti," *Zoologica,* vol. 10, no. 1 (1928), pp. 1–279; and "Additions to the Fish Fauna of Haiti and Santo Domingo," ibid., vol. 10, no. 4 (1935), pp. 317–319.

6. Oliver L. Austin, Jr., *Birds of the World* (New York, Golden Press, 1961), p. 168. A total of 687 species and subspecies, with nineteen doubtfully distinct or uncertain, are listed under *Trochilidae* by James Lee Peters, *Check-List of Birds of the World* (Cambridge, Mass., Harvard University Press) vol. 5 (1945), pp. 3–143.

7. *The Log of the Sun,* p. 117.

8. In the early 1930s Beebe and his staff moved to the Biological Station and New Nonsuch Laboratories a short distance west of Saint Georges.

9. William Beebe and John Tee-Van, *Field Book of the Shore Fishes of Bermuda.* New York, G. P. Putnam's Sons, 1933.

10. *Nonsuch: Land of Water,* p. 126.

11. Against this German triumph America can pose a modest but authentic claim of priority, with the first wartime sinking of any ship by a type of submarine. Early in 1864 the Confederate submersible *David* ran a spar torpedo into the side of the U.S.S. *Housatonic* in Charleston harbor, sending that ironclad to the bottom and going down simultaneously with all hands.

12. See James Dugan, *Man Under the Sea* (New York, Harper & Bros., 1956), pp. 138–145, p. 311 (first entry). This work is, according to Jacques-Yves Cousteau, "the standard history" of the subject.

13. See especially the *New York Times,* November 25, 1926, p. 1.

14. Otis Barton, *The World Beneath the Sea* (New York, Thomas Y. Crowell Co., 1953), p. 7. The first six chapters of this cheerfully quixotic book deal *passim* with the bathysphere story.

15. Ibid., p. 12.

16. Briefly, the two photos facing p. 152 in *Half Mile Down* were posed some time after the actual event. Water coming from the sphere is not spurting out under pressure, but merely splashing to the deck, and it is emerging not from the wing bolt (as alleged) but from the door itself, with all retaining nuts removed (cf. upper photo facing p. 170). Also, on p. 154 Beebe said "two of us" turned the wing bolt, but here he is bravely doing the job himself, with Barton (left) singularly inattentive, or perhaps unhappy with the charade.

17. With proper scientific caution, however, Beebe admits that this fish "may belong in [a] quite different or wholly unknown [family]"—a qualification which may also be applied to the Five-lined Constellation Fish, *Bathysidus pentagrammus,* and perhaps to the Pallid Sailfin, *Bathyembryx istiophasma,* and the Three-starred Anglerfish, *Bathyceratias trilynchnus*—all named from no direct evidence other than a sighting or two from the bathysphere window. See *Half Mile Down,* pp. 172–173, 205, 211–212, 312, 315, 323, 327.

18. See preceding note.

19. *The Arcturus Adventure,* p. 382.

20. William Beebe, "A Round Trip to Davy Jones's Locker," *National Geographic Magazine,* June 1931, Plate V. The same plate appears in *Half Mile Down* facing p. 218, but with the binomial changed.

21. Dugan, *Man Under the Sea,* pp. 295–306; also Jacques-Yves Cousteau, with James Dugan, *The Living Sea* (New York, Harper & Row, 1953), *passim.* In January 1960 a bathyscaphe reached an ultimate depth of nearly 36,000 feet in the Marianas Trench, and of late several independent depth vehicles have been developed, for example Cousteau's nimble "diving saucer," Jacques Piccard's *Ben Franklin,* the U.S. Navy's Deep Submergence Rescue Vehicle, the Reynolds *Aluminaut,* and others.

22. William Beebe, "Oceanographic Work at Bermuda of the New York Zoological Society," *Science,* November 30, 1934, p. 495. This brief and sober report is especially interesting for the scientific inferences Beebe makes from his diving experiences.

23. Jacques-Yves Cousteau and James Dugan, eds., *Captain Cousteau's Underwater Treasury* (London, Hamish Hamilton, 1960), p. 281.

24. Frank M. Chapman, *Autobiography of a Bird-Lover* (New York, D. Appleton–Century Company, 1933), pp. 71–72.

25. "Reviews and Comments," *Copeia: A Journal of Cold Blooded Vertebrates,* July 16, 1935, p. 105.

26. *Natural History: The Journal of the American Museum of Natural History,* January 1935, pp. 88–89.

27. Rachel L. Carson, *The Sea Around Us* (New York, Oxford University Press, 1951) p. vi.

8. FEMALES OF THE SPECIES

1. *Galápagos: World's End,* p. 326.

2. See Gloria Hollister, "Clearing and Dyeing Fish for Bone Study," *Zoologica,* vol. 12, no. 10 (1934), pp. 89–101.

3. William Beebe, *Zaca Venture.* New York, Harcourt, Brace and Co., 1938.

4. *Galápagos: World's End,* p. 29.

5. That Beebe in his professional relationships was also a willful and often exasperating man—especially to William T. Hornaday, the autocratic director of the Zoological Park from 1896 to 1926—is equally clear. See William Bridges, *Gathering of Animals: An Unconventional History of the New York Zoological Society* (New York, Harper and Row, 1974), pp. 124–125, 292–294, 297–304, 365–366, 389–391, 394–397, 427–428, 437–438.

6. William Beebe, "The Jelly-Fish and Equal Suffrage," *Atlantic Monthly,* July 1914, pp. 36–47.

7. Elswyth Thane, *Reluctant Farmer.* New York, Duell, Sloan and Pearce, 1950. Miss Thane's *The Bird Who Made Good* (1947) also concerns the Vermont place.

9. HOME FROM THE SEA

1. *High Jungle,* p. 49.

2. Ibid., p. 203.

3. See G. A. C. Herklots, *The Birds of Trinidad and Tobago* (London, Collins, 1961); and Malcom Barcant, *Butterflies of Trinidad and Tobago* (London, Collins, 1970).

4. Jocelyn Crane, "Keeping House for Tropical Butterflies," *National Geographic Magazine,* August 1957, pp. 193–217. This and the aforementioned Beebe article (June 1958) are useful both for data and for photos of Crane and Beebe and the Simla station.

5. William Beebe, *Unseen Life of New York As a Naturalist Sees It.* New York, Duell, Sloan and Pearce; Boston, Little, Brown and Co., 1953.

6. The twenty-one titles include books written jointly with Mary Beebe, G. Inness Hartley and Paul G. Howes, and John Tee-Van, plus the anthology *The Book of Naturalists* (New York, Alfred Knopf, 1944), which Beebe edited with warmly acknowledged help from Jocelyn Crane. Not included are two collections from his own works, *Exploring With Beebe: Selections for Young Readers* (New York, G. P. Putnam's Sons, 1932) and *Adventuring With Beebe: Selections from the Writings of William Beebe* (Boston, Little, Brown, and Co., 1955).

7. *The Book of Naturalists,* p. 73.

8. *Book of Bays,* p. 74.

9. Letter from A. E. Hill, M.D., F.R.C.P.(C), to the author, December 6, 1971.

10. Ibid.

11. In 1974 Simla was acquired by the Asa Wright Nature Centre, a sanctuary and research station located some four miles up the Arima valley.

10. MAN OF SCIENCE

1. William Beebe, "The Hills," *The Atlantic Monthly,* June 1916, p. 778.

2. Gerhard Heilmann, *The Origin of Birds* (London, H. F. & G. Witherby, 1926), pp. 194–199.

3. Jean Delacour, *The Pheasants of the World.* London, Country Life, Ltd.; New York, Charles Scribner's Sons, 1951.

4. Although Beebe is not accounted a successful theorist, flat statements of failure should be avoided in evolutionary matters. Archeopteryx remains paleontologically isolated, but the very real chance remains that an earlier fossil form may one day come to light, perhaps not Tetrapteryx, but perhaps not Heilmann's "Proavis" either.

5. Quoted in *Half Mile Down,* pp. xi–xii.

6. Interview, April 2, 1971.

7. "Sed quis custodiet ipsos custodes?" Juvenal, *Satire VI,* 347.

8. Eldon J. Gardner, *History of Biology* (Minneapolis, Burgess Publishing Co., 1965), p. 351.

9. Besides his honorary doctorates, Beebe received the Daniel Giraud Elliot Medal of the National Academy of Science in 1918, the Geoffrey St. Hilaire Medal from the Société d'Acclimitation de France in 1921, the first Burroughs Medal of the John Burroughs Memorial Association in 1926, and the Theodore Roosevelt Distinguished Service Medal from the Theodore Roosevelt Association in 1953. Beebe was often written up in newspapers and magazines, and was invited to appear as guest expert on NBC's widely popular "Information, Please" program in 1939, 1941, and 1943.

10. *The Log of the Sun,* p. 305.

11. *The Bird,* p. 482.

12. William Beebe, "Ecology of the Hoatzin," *Zoologica,* vol. 1, no. 2 (1909), p. 63.

13. *Unseen Life of New York as a Naturalist Sees It,* p. 9.

14. Quoted in C. Brooke Worth, *Naturalist in Trinidad* (Philadelphia, J. B. Lippincott Co., 1967), p. 120.

15. See Helen Hays and Robert W. Risebrough, "The Early Warning of the Terns," *Natural History,* November 1971, pp. 39–46.

16. *The Log of the Sun,* pp. 10–11.

17. A pertinent essay (aptly titled "What Is Ecology?") may be found in Philip Handler, ed., *Biology and the Future of Man* (New York, Oxford University Press, 1970), pp. 431–473.

18. Konrad Lorenz, *King Solomon's Ring* (London, Methuen & Co., 1964), pp. vii, xv.

19. In October 1973 Lorenz, Tinbergen, and the Vienna-born biologist Karl von Frisch were awarded the Nobel Prize for their distinguished work in particular areas of ethology.

20. *Nonsuch: Land of Water,* p. 138.

21. Ibid., p. 252.

22. George B. Schaller, *The Year of the Gorilla.* Chicago, The University of Chicago Press, 1964. A year earlier the same publisher brought out Schaller's *The Mountain Gorilla,* documenting gorilla behavior in great detail. More recently Schaller has published *The Deer and the Tiger* (University of Chicago Press, 1967) and *The Serengeti Lion* (1972), the latter a

volume of the University of Chicago Press series *Wildlife Behavior and Ecology,* with Schaller serving as editor.

23. Jane van Lawick–Goodall, *In the Shadow of Man.* Boston and New York, Houghton Mifflin Company, 1971. *Innocent Killers,* a study of hyenas, jackals, and wild dogs written jointly by Hugo and Jane van Lawick–Goodall, was published by Houghton Mifflin the same year.

11. THE FINALITY OF WORDS

1. *The Book of Naturalists,* p. 87.
2. *New York Times Book Review,* October 1, 1922, p. 3.
3. Ibid.
4. Recently much quoted, this assertion first appeared in "Walking," the lead article in *The Atlantic Monthly* for June 1862, p. 668.
5. George Reuben Potter, "William Beebe: His Significance to Literature," *University of California Publications in English,* vol. 1, (1929), p. 222.
6. *Half Mile Down,* p. 196.
7. John Steinbeck and Edward F. Ricketts, *Sea of Cortez.* New York, Viking Press, 1941. In the "Annotated Phyletic Catalogue and Bibliography" (pp. 283–589) Beebe is cited for several *Zoologica* articles, *Galápagos: World's End,* and *Zaca Venture,* and Jocelyn Crane for *Zoologica* papers on crabs.
8. John Steinbeck, *The Log from the Sea of Cortez* (1951). London, Pan Books Ltd., 1960.
9. Ibid., pp. 147, 267.
10. For a detailed analysis of the stories cited, see Richard B. Hovey, *Hemingway: The Inward Terrain* (Seattle and London, The University of Washington Press, 1968), pp. 6–8, 32–36.
11. William Beebe, "A Red Indian Day," *The Atlantic Monthly,* July 1918, pp. 23–31.
12. Ibid., pp. 24–25.
13. *Pheasant Jungles,* pp. 100–101.
14. Ibid., p. 107.
15. Ibid., p. 207.

Books by William Beebe

Two Bird-Lovers in Mexico
HOUGHTON MIFFLIN CO., 1905

The Bird, Its Form and Function
HENRY HOLT AND CO., 1906

The Log of the Sun
HENRY HOLT AND CO., 1906

Our Search for a Wilderness (with Mary Blair Beebe)
HENRY HOLT AND CO., 1910

Tropical Wild Life in British Guiana (with G. Inness Hartley and Paul G. Howes)
NEW YORK ZOOLOGICAL SOCIETY, 1917

Jungle Peace
HENRY HOLT AND CO., 1918

Edge of the Jungle
HENRY HOLT AND CO., 1921

A Monograph of the Pheasants (4 vols.)
H. F. WITHERBY AND CO., 1918–1922

Galapagos: World's End
G. P. PUTNAM'S SONS, 1924

Jungle Days
G. P. PUTNAM'S SONS, 1925

The Arcturus Adventure
G. P. PUTNAM'S SONS, 1926

Pheasants: Their Lives and Homes (2 vols.)
DOUBLEDAY, PAGE AND CO., 1926

Pheasant Jungles
G. P. PUTNAM'S SONS, 1927

Beneath Tropic Seas
G. P. PUTNAM'S SONS, 1928

Nonsuch: Land of Water
BREWER, WARREN, AND PUTNAM, 1932; HARCOURT, BRACE AND CO., 1932

Exploring with Beebe (selections)
G. P. PUTNAM'S SONS, 1932

Field Book of the Shore Fishes of Bermuda (with John Tee-Van)
G. P. PUTNAM'S SONS, 1933

Half Mile Down
HARCOURT, BRACE AND CO., 1934

Zaca Venture
HARCOURT, BRACE AND CO., 1938

Book of Bays
HARCOURT, BRACE AND CO., 1942

The Book of Naturalists (edited)
ALFRED A. KNOPF, INC., 1944

High Jungle
DUELL, SLOAN AND PEARCE, 1949

Unseen Life of New York as a Naturalist Sees It
DUELL, SLOAN AND PEARCE, 1953; LITTLE BROWN AND CO., 1953

Adventuring with Beebe (selections)
LITTLE, BROWN AND CO., 1955

Index